THERMAL AND ENERGETIC STUDIES OF CELLULAR BIOLOGICAL SYSTEMS

THERMAL AND ENERGETIC STUDIES OF CELLULAR BIOLOGICAL SYSTEMS

Thermal and Energetic Studies of Cellular Biological Systems

Edited by

A. M. James, MA, DPhil, DSc, CChem, FRSC

Professor Emeritus, University of London

WRIGHT
Bristol 1987

Published under the Wright imprint by
IOP Publishing Limited
Techno House, Redcliffe Way, Bristol BS1 6NX

British Library Cataloguing in Publication Data

Thermal and energetic studies of cellular biological systems.
1. Calorimeters and calorimetry 2. Biology—Technique
I. James, A.M. (Arthur Montague)
574.19′285 QH324.9.C3

ISBN 0 7236 0909 8

Typeset by
Datapage International Ltd., Dublin, Ireland

Printed in Great Britain by The Bath Press, Lower Bristol Road, Bath BA2 3BL.

Preface

This book, a review of the literature on the thermal and energetic changes which occur in living organisms, was commissioned at the suggestion of Dr Philip Edge of John Wright & Sons. It seeks to emphasize the importance of information obtained from non-destructive calorimetric techniques in increasing our knowledge of biological processes. The book is a natural successor to the more broadly based book edited by Dr Beezer (*Biological Microcalorimetry*, Academic Press, 1980).

Thermal studies are making an important contribution to our understanding of the properties of a wide range of cellular systems and non-cellular systems (e.g. biomembranes and their molecular components, model systems and enzymes). However, it was considered that, following an international symposium, there was sufficient interest and available research material to restrict this volume to a discussion of energy changes in cellular systems.

Even so it has not been possible, within the limits of a book of this size, to cover in detail all the possible cellular systems and their interactions in great depth. Each author has therefore presented a concise, up-to-date account of his field of expertise. These accounts, it is hoped, will be of help to workers seeking a new non-destructive technique for the study of a cellular system, will be a useful guide to the novice and a handy reference book for those already active in the field.

The selection of the topics is entirely the responsibility of the Editor and has been formulated as a result of discussions with colleagues and research students. The Editor is grateful for the cooperation of the contributors in the production of their manuscripts in the allotted time and to Dr Janet Bunker for her helpful discussions and comments during the preparation of his own contributions.

A. M. J.

Contributors

Anthony E. Beezer. Reader in Biophysical Chemistry, Royal Holloway and Bedford New College, Egham Hill, Egham, Surrey TW20 0EX, UK.

Larry E. Erickson. Professor of Chemical Engineering, Kansas State University, College of Engineering, Durland Hall, Manhattan, Kansas 66506, USA.

Arthur M. James. Professor Emeritus of University of London, Royal Holloway and Bedford New College, Egham Hill, Egham, Surrey TW20 0EX, UK.

Richard B. Kemp. Reader in Zoology, The University College of Wales, Penglais, Aberystwyth, Dyfed SY23 3DA, UK.

Roger J. Miles. Lecturer in Microbiology, King's College (KQC), Campden Hill Road, London W8, UK.

Mario Monti. Associate Professor, Department of Internal Medicine, University Hospital of Lund, S-221 85, Lund, Sweden.

Brian F. Perry. Senior Scientist, Richardson-Vicks International, Rusham Park, Whitehall Lane, Egham, Surrey TW20 9NW, UK.

Ingemar Wadsö. Professor of Thermochemistry, University of Lund, Chemical Center, PO Box 740, S-220 07 Lund, Sweden.

John Widdows. Principal Scientific Officer, Institute for Marine Environmental Research, Prospect Place, The Hoe, Plymouth, Devon PL1 3DH, UK.

Contents

Chapter 1

Calorimetry, Past, Present, Future

A. M. James

The production of heat by living organisms is one of the transformations of energy brought about by these cells. In everyday life this is manifested by the rise in temperature of fermenting liquids, decomposing manure and other organic material (e.g. straw, cellulose). The heat evolved during the bacterial degradation of organic substances can have disastrous consequences, for example in the spontaneous combustion of refuse heaps as a result of overheating. During the production of organic solvents, antibiotics, etc. by fermentation much heat is generated; this has to be removed. Thus any information relating to the amount of heat energy produced and its rate of production is important.

Calorimetry is a non-specific technique for the direct measurement of metabolic activity. It is a measure of the algebraic sum of the enthalpy changes of all the biological processes in the cell, even if the nature of these processes is not fully understood. To increase specificity and increase the understanding of the complex biological reactions other parameters of the system must be monitored. In some instances the non-specific nature of the method is advantageous in the study of metabolic processes. For example, changes in metabolism as a result of changed envoronmental conditions, revealed by thermal changes, can often be detected at lower cell populations than by conventional techniques.

1.1 Historical

The first recorded calorimetric experiment on microorganisms concerned the heat of fermentation of yeast developed in a large brewer's vat (21 400 l)

(Dubrunfaut 1856); this cannot be considered as microcalorimetry! Rubner (1908) was the first to undertake the first true calorimetric measurements during the fermentation and growth of yeast. Prior to this values of the energy of metabolism were evaluated from measurements of the heat content (enthalpy) of the nutrient medium before and after growth of the organisms (and sometimes the heat content of the biomass) in a bomb calorimeter. The heat evolved during growth was calculated using Hess's law of constant heat summation (Rubner 1903).

The development of a differential calorimeter (Hill 1912; Bayne-Jones 1929) permitted the first study of energy changes during bacterial growth and the determination of the rate of cellular catabolic activity. Bayne-Jones and Rhees (1929) reported that:'Changes in the rate of heat production are related to changes in the growth rate of bacteria. Young bacterial cells produce more heat per cell than older ones.' These authors calculated that the heat produced for growing cells of *Escherichia coli* was in the range 4–40 nJ cell^{-1}; the actual value depending on the composition of the growth medium. Shearer (1921) observed that bacteria liberate much less heat when utilizing free amino acids than when they use more complex protein split products. In her very extensive study of thermal events during growth of several organisms, Bunker (1985) observed that more heat was produced during growth in simple defined medium than in rich nutrient medium. In defined media the organisms have to use some of the energy from the carbon source to synthesize essential metabolites which are present in rich media (p. 87).

The advent of sensitive heat conduction microcalorimeters, reviewed by Wadsö (p. 34) permitted more detailed observation of thermal events which occur during the growth of bacteria (p. 75) and yeasts (p. 108) in simple and complex media. This has increased our understanding of bacterial growth kinetics as well as giving information on enthalpy changes during the growth cycle.

There are many advantages to using the technique of microcalorimetry to study biological systems; the measurements are reproducible, rapidly obtained and the results readily related to biomass or metabolic rates. Further, in contrast to other techniques, the method does not depend on the use of homogeneous media.

1.2 Reporting Experimental Results

Although heat conduction calorimeters operate with a sensitivity of 100 μV mW^{-1} and a resolution of better than 1 μV (i.e. heat flows between 1 and 10 μW) they are far too insensitive to register energy changes associated with one cell. At least 10^5 cells, and preferably more, each with a heat production of about 10 pW, are thus necessary to obtain a signal significantly greater than the noise level. This means that cultures or suspensions with a minimum population of 10^5–10^6 cells cm^{-3} are required for study. Enthalpy changes or power expressed per cell are therefore derived from the average value for a large population and a knowledge of its size; no measured values are available for an individual cell.

Instruments with an increased sensitivity capable of detecting energy changes in populations as low as 10^3 cells cm^{-3} would be extremely valuable; it would then be possible to observe thermal events which occur during the lag and early growth phases of cultures in the presence and absence of antibacterial agents, and permit measurement of the maintenance energy of non-dividing cells in the stationary phase of the cell cycle and possible correlation with the size of ATP pool of the cells. It would also permit the study of suspensions of cells which are only available in small quantities (e.g. tumour cells isolated from patients) and possibly provide an alternative and continuous method of monitoring cell numbers.

Experimental data from the calorimeter is presented as a heat flow or power output; the value varies with the type and age of the cells under observation and the nature of the suspension medium. Typical values for a selected range of cells are given in Table 1.1: the values for bacteria and yeasts are for growing cultures, the remainder are for suspensions of resting cells. Normally the data is in the form of a power-time (p-t) trace, formerly known as a 'thermogram' which can be recorded as a trace or stored for future handling of the data in a microprocessor. The shape and position of the trace depends on such factors as the nature of the cell (bacterial, yeast, blood, tissue cell); the composition of the growth or suspension medium including the presence or absence of inhibitors; aerobic or anaerobic conditions of test and the nature of the calorimeter (batch or flow).

During the exponential growth of microorganisms the power increases exponentially, usually with a rate constant similar to that of biomass production. If growth is nutrient-limited then there is a marked decrease in the power to zero when the nutrient is exhausted (Figure 1.1). In richer nutrient medium the power output in the stationary phase of growth never becomes zero on account of the energy released during the metabolism of secondary intermediates present in the medium.

It is surprising that so little experimental data is available for microorganisms growing in continuous or chemostat culture. This deficiency is probably due to difficulties arising from the need to pump samples from the chemostat to the calorimetric vessel which limits the range of dilution rates (and hence growth rates) and possibly results in the destruction of the steady state condition. The recent development of an instrument (Leiseifer and Schleser 1982) in which the fermentor also functions as the calorimeter vessel may overcome these problems. The advantages of studying such cultures is that the cells are reproducing in a constant environment and the heat output is invariant with time; changes brought about by the presence of inhibitors can therefore be monitored.

Table 1.1 Power output for a selected range of biological cells

Species		Heat production (mean values) /pW cell^{-1}	/W g^{-1}
Bacteria (Chapter 4)			
Klebsiella aerogenes			
(continuous culture, glucose)		0·5	0·4
(glucose-limited, batch)	max.	4–5	3·2–4
	exp.	2–2·5	1·6–2
(nutrient broth)	max.	1·5	1·2
Staphylococcus aureus			
(nutrient broth)	max.	1·5–2	1·2–1·6
Yeasts (Chapter 5)			
Saccharomyces cerevisiae			
(aerobic, complex medium)			2·15
(anaerobic, complex medium)			0·34
Fungi			
Fusarium roseum		40	
Human blood cells (Chapter 6)			
Erythrocytes		10×10^{-3}	
Platelets		60×10^{-3}	
Lymphocytes		2	
Granulocytes (phosphate buffer pH 9)		1·3	
(plasma)		3·5	
Tissues (Chapter 8)			
Human skin fibroblasts		40	
Keratinocytes		80	
Adipocytes (normal)		49	
(obese)		26	
Aquatic animals (Chapter 9)			
Various (range)			$0{\cdot}08–3 \times 10^{-3}$
Fish eggs			$0{\cdot}25 \times 10^{-3}$

Apart from considerable inprovement in the sensitivity and design of calorimeters and calorimeter vessels, perhaps the most important change which has occurred in the past 10 years is the realization that *p-t* traces on their own have little significance, and that thermal data must be supplemented with other chemical and/or biological data. This is emphasized by each of the contributors. As a result, it is therefore possible to make quantitative calculation of enthalpy changes. Information on the variation of such parameters as

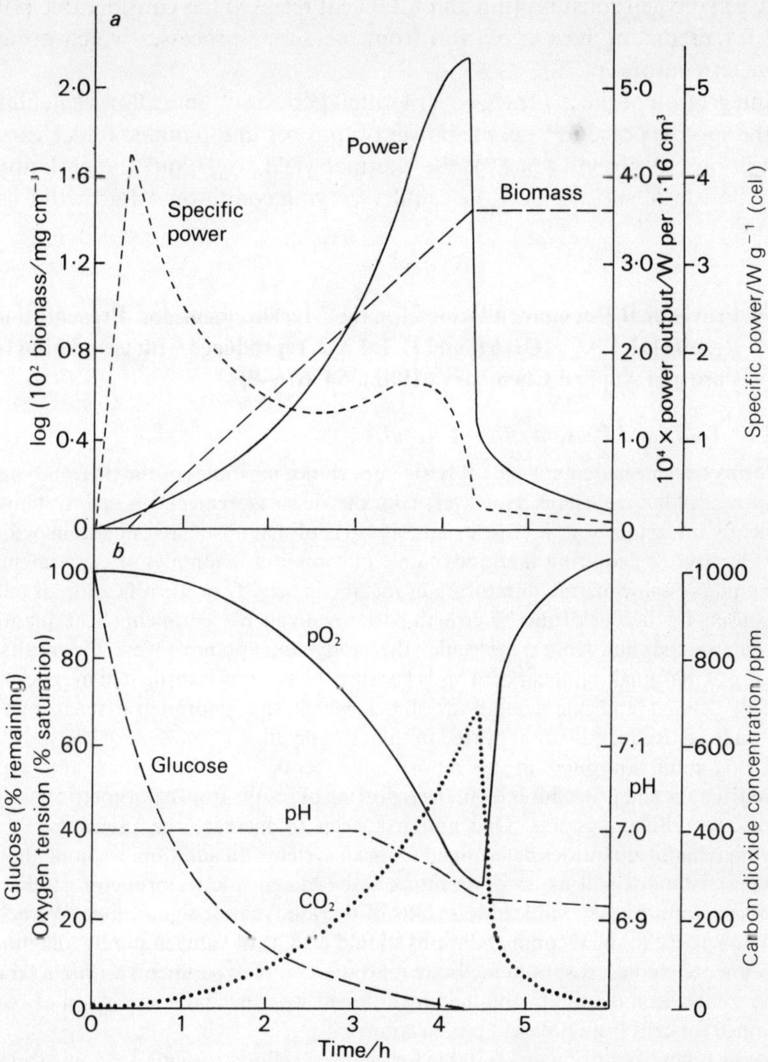

Figure 1.1. Changes in the power and other growth parameters during the growth of *Klebsiella aerogenes* (NCTC 418) in glucose-limited medium (3·3 mmol dm^{-3}) at 37 °C. Flow rate through the calorimeter vessel = 90 cm^3 h^{-1} (After Nichols 1980). *a* Power; specific power; biomass. *b* Oxygen; glucose; carbon dioxide; pH.

biomass production, yield, oxygen consumption, carbon dioxide evolution, size of the ATP pool and chemical composition of the growth medium are now commonly reported. With such data it is possible to establish qualitative and quantitative relationships between the various parameters, e.g. the oxycaloric equivalents in aerobic metabolism (p. 192). The established correlation between oxygen consumption and total heat released has considerable potential for predicting heat evolution from metabolic processes which occur in large fermentors (p. 79).

Integration of the *p-t* traces over limited periods of time allows calculation of the specific power, *P*, i.e. the power output per unit biomass under varying conditions of growth (p. 80), the thermal yield, $\Delta_p H$ and the enthalpy of metabolism of substrates $\Delta_{met} H$, under varying conditions of growth (Table 4.1).

1.3 Provisional Recommendations for the Measurement and Presentation of Results [IUPAC, IUPAP and IUB 1982, reproduced with permission from Pure and Applied Chemistry (1982), 54, 676–9]

1.3.1 Units and Presentation of Results

Calorimetric measurements are of basic importance in studies of the thermodynamic properties of living systems. However, calorimetric measurements on such systems are presently undertaken, to a large extent, as parts of diagnostic investigations without the objective of acquiring thermodynamic information. Examples of experiments of this kind are calorimetric monitoring of metabolic activity or identification of microorganisms by their calorimetric growth patterns. In such experiments heat quantities are determined, but, from a molecular thermodynamic point of view, the results are usually of marginal significance. This is because the systems investigated in general are poorly defined and the conditions under which the calorimetric measurements are made are frequently not reported in sufficient detail.

The Commission offers in this report some recommendations concerning experimental details and procedures from the reporting of results from calorimetric measurements on cellular systems. One aim has been to increase the usefulness of the thermodynamic quantities determined for such systems. In addition, we hope that the recommendations will assist in orienting some 'diagnostic' calorimetric studies towards procedures that will lead to results of thermodynamic significance. We believe that adherence to the recommendations should also be of value in purely 'diagnostic' calorimetric studies. It should facilitate reproduction of experiments within a laboratory, comparison of results obtained by different workers and comparison of results obtained for cells from different preparations.

Many types of calorimeters suitable for work on cellular systems are now available. The measured heats are often small and 'microcalorimeters' are normally employed. The term 'microcalorimeter' is not strictly defined but in current practice it usually implies (for reaction calorimeters) that the smallest measurable thermal power is of the order of 1 μW and that the reaction vessel is small, of the order of a few cm^3 or less.

There is a general and lasting value in well documented energy data for biological systems. Thus, the use of arbitrary units like 'mm recorder deflection' should be avoided. Instead the calorimeter should be calibrated and the results reported in SI units. Calorimeters are normally calibrated electrically. The measurement of electrical energy is, today, a trivial procedure that can easily be made with an accuracy exceeding the needs of biological experiments. The problem is rather to make certain that the electrical energy is released in a manner closely comparable with that for the process studied. Microcalorimeters used in biological work are not always well suited for such strict comparisons. In particular, for flow-through vessels calibration procedures are often far from ideal, and it is currently rare that the results are checked by a suitable test reaction.

The response time of a calorimeter can be significant compared to the duration of the process being studied. Therefore, the calorimetric signal at any point in time may not have a simple relationship with the thermal power generated at that time within the reaction vessel. This must be taken into account when kinetic information is deduced from the results of calorimetric experiments.

It is necessary to specify carefully the physiological conditions for the cellular material during a calorimetric measurement. Special attention must be given to the practical design of the reaction vessel, the performance of the experiment and to the methods by which the physiological conditions in the reaction vessel are verified.

Presentation of Experimental and Derived Results

The output of many calorimetric experiments is a plot of power (energy evolution per unit of time) as a function of time. The recommended unit for power (P) is watt (W) and for time the second (s) or some multiple thereof [e.g. kilosecond (ks)]. With the use of these units the time integral of the power-curve is given directly in joules (J) or some multiple thereof.

In graphs the axes should be labelled with both the name (or the symbol) of the quantity and the unit employed. Preferably the label should show the quantity divided by the unit in order to emphasize that the scale markers represent pure (dimensionless) numbers.

1.3.2 Some Specific Recommendations

The Cell Preparation

Where possible, experiments should employ material obtained from type collections with well defined organisms or cells. In other cases detailed information about the origin and method of preparation (isolation) should be reported. For cells which have been stored, maintenance conditions such as time, temperature and media should be reported. Details concerning passages of the organisms from the state of storage to experimental use should also be described (number of subcultures and details of the growth conditions). Counts of contaminating cells, e.g. for blood cell preparations, should be reported.

The Calorimeter

Full details of the design of the calorimeter and its performance should be reported. Alternatively, reference can be given to other reports provided these are readily available. However, even minor changes in apparatus design and experimental procedure should be indicated. In each report the following details require attention:

(1) The design of the calorimetric vessel, its volume and its construction material, gas phase, method for initiation of a process, method used for stirring or agitation of the liquid medium.

(2) Time constants of the apparatus.
(3) Results of calibration and test experiments including accuracy assignment.
(4) Base line stability (where relevant) over a period of time which normally should be greater than the measurement period.
(5) Amplification and recording equipment; pumps and other ancillary equipment.
(6) Method employed for cleaning and sterilization of the calorimetric vessel and flow lines where appropriate.

The Calorimetric Experiments

Experimentalists are reminded that all processes, both physical and chemical, will contribute to the observed heats. Thus, processes such as dilution, protonization, mixing, vaporization, etc. may give rise to significant systematic errors if not taken into account in the experimental design and in the evaluation of the results. In the special case of twin calorimeters, where two processes are compared, it is frequently useful to observe the separate output from each vessel to aid in the interpretation of the difference signal.

In each study the following details require attention:

(1) The measurement temperature must be reported.
(2) It is suggested that, where possible, chemically defined media should be used. Such media can be accurately reproduced and a detailed chemical analysis of reaction mixtures will be facilitated. In any case full details about the medium composition must be reported.
(3) If a gas or a liquid is perfused into, or through, the reaction vessel this must be reported in detail.
(4) Full details concerning inoculation of the medium must be reported (size, age and storage condition of inoculum, medium used and, where relevant, the growth phase). For preparations of, for instance, blood cells, counts of contaminating cells should be made and reported.

    For growth experiments (microorganisms, cultured tissue cells) information about changes in the cell count is often essential. In any case initial and final cell concentrations (dry mass, etc.) should normally be reported, together with results of viability tests.

    Details of the enumeration method followed, together with an estimate of the accuracy of the count, should be given. Relevant details of the extent of damaged cells should be given. (Note: intact enzyme substrate systems may contribute to the observed thermal power.)
(5) For the evaluation of results of calorimetric experiments, and to ensure their lasting value, heat measurements should ideally be accompanied by detailed analysis (pH, concentration of O_2, CO_2, energy source(s), metabolites, etc.) performed continuously, or at intervals, on the reaction medium.

    For almost all calorimetric experiments on cellular material it is essential that initial and final pH values be reported and that information be given concerning the concentration of oxygen. The problem with concentration gradients in experiments performed under static conditions should be noted, cf. effects of cell sedimentation.

Sometimes the necessary analytical information (including cell counts) must be obtained from parallel non-calorimetric experiments. It is important to perform such experiments under conditions which are as close as possible to those in the calorimetric experiment. They should be performed simultaneously, there should be no scale effect, vessels should be of the same material, medium and gassing rates, etc. should be identical. Even if great care is taken in performing such parallel experiments the cellular processes may not proceed exactly as expected. This must be taken into account when uncertainty assignments are made.

(6) Experiments with cells are often carried out in suspensions. Depending on the type of organism, and on the calorimetric method, the cells may stay in a uniform suspension or they may partly or completely sediment before, or during, the calorimetric experiment. Cells originally in suspension may adhere to the walls of the calorimetric vessels. In flow calorimetric experiments, therefore, it is sometimes difficult to assess accurately the number of cells contained in the reaction vessel (during the experiment or at the end of the experiment) as they may have been partially trapped in the flow lines or they may have accumulated in the reaction vessel. Effects of this nature can be of crucial importance for the evaluation of the calorimetric results. It is, therefore, most important to consider these effects in the design of the calorimetric experiments and in the reporting of the results.

(7) When using differential scanning calorimeters for the study of intact cellular systems attention must be paid to the scanning rates employed. The scanning rates used should attempt to ensure equilibrium conditions throughout the temperature range of interest. For comparative purposes it is advisable to extrapolate the recorded data to zero heating velocities. Thus, experiments should be conducted using a minimum of three different heating rates.

The use of the term power-time curve is recommended in place of the commonly used term 'thermogram' to avoid confusion with, for example, its use in differential thermal analysis.

It is often useful to derive values for the thermal power produced per cell, for a defined mass (dry weight) of cells or for a defined volume (wet, packed) of cells. Example: For freshly prepared human erythrocytes in plasma an approximate power per cell at 37 °C and pH 7·40 is $P = 8\,\mathrm{fW}$. For wet packed cells this value corresponds to about $80\,\mathrm{mW\,dm^{-3}}$. Alternatively it may be of interest to report heat quantities associated with the consumption of specified substrates. When such values are reported they must be accompanied by relevant information about the experimental conditions since the power or heat quantities evolved can be very sensitive to variations in experimental conditions such as pH, medium composition (including concentrations of oxygen and carbon dioxide), temperature, storage conditions for the cells (time, temperature), cell concentration, etc.

The power-time curve reported should take into account the dynamic characteristics of both the calorimeter and the process under study. Two situations can arise:

(1) Where the dynamic parameters of the calorimeter do not affect the experimental power-time curve (e.g. where steady state processes are observed or where the time constant for the process studied is long compared to that of the instrument),

(2) Where the experimental power-time-curve requires correction to take account of these instrumental parameters.

The corrections under (2) should be expressed in sufficient detail to permit regeneration, if required, of the experimental power-time-curve.

Where instrumental data handling systems take into account these dynamic characteristics prior to presentation of results, the appropriate references or descriptions of the method of correction employed and the values of the defined parameters should be reported.

Interpretation of Observed Heat Quantities Associated with Growth Processes

If possible, the enthalpy change associated with growth phenomena should be expressed in thermodynamic terms. It is recommended that the enthalpy change per unit of substrate metabolized be reported, ΔH_{met}.* The recommended unit for ΔH_{met} is $J\,mol^{-1}$. When the catabolic process is well known it is possible to calculate the enthalpy change corresponding to catabolism, ΔH_{cat}. The enthalpy change corresponding to anabolism is similarly expressed as ΔH_{an}. Thus, in well defined systems when an energy substrate is metabolized, the enthalpy balance can be written:

$$\Delta H_{met} = (1 - \alpha)\Delta H_{cat} + \alpha \Delta H_{an} \tag{1-1}$$

α represents the fraction of the carbon source which is incorporated into the cellular material and can be calculated from the molecular growth yield.

When the nitrogen source is simple (e.g. NH_3, NO_3^-, N_2) it is possible to calculate the enthalpy change corresponding to anabolism from values for the elemental composition of the cellular material, enthalpies of combustion of the dried cells and the molecular growth yields. For this calculation it is simpler to express the results as the specific enthalpy change, i.e. the enthalpy change per unit mass of dried cells, Δh_{an}.

$$\Delta h_{an} = \frac{\alpha \Delta H_{an}}{Y_s}$$

where Y_s is the molecular growth yield. Recommended units are for Δh_{an}: $J\,g^{-1}$ and for Y_s: $g\,mol^{-1}$. Equation (1.1) then becomes

$$\Delta H_{met} = (1 - \alpha)\Delta H_{cat} + Y_s \Delta h_{an}.$$

[Pure and Applied Chemistry (1982), 54, 676–9]

1.3.3 Examples of Presentation of Results

The most serious problems are concerned with the reporting of thermal changes during the growth of bacteria and yeasts where insufficient relevant

---

*It is recommended that symbols should be written $\Delta_{met}H$ and $\Delta_{cat}H$, etc. rather than ΔH_{met}, ΔH_{cat}, etc.

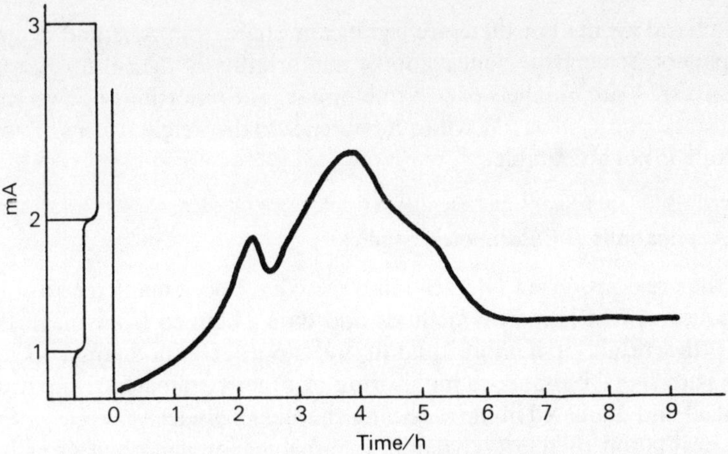

Figure 1.2. Microcalorimeter response during the growth of *Staphylococcus aureus* ham 13·655 in Columbia broth at 35 °C. Flow rate through the calorimeter vessel = 25 cm^3 h^{-1} (After Semenitz 1979).

information is reported. Results of thermal studies on resting cells normally contain sufficient information about the mass or number of cells or weight of tissue, the nature of the suspension medium and the presence of added substrates or inhibitors.

Two typical (possibly extreme) examples of presentation of results obtained during the growth of organisms are shown in Figures 1.1 and 1.2. For the growth of a strain of *Klebsiella aerogenes* obtained from the National Collection of Type Cultures, experimental details of the culture medium, growth and environmental conditions and of the thermal measurements are clearly stated (Nichols 1980). The *p-t* trace (Figure 1.1) can be interpreted in terms of the variation of other growth parameters and in addition meaningful thermodynamic data ($\Delta_p H$ and $\Delta_{met} H$) can be evaluated.

In contrast (Figure 1.2), for the growth of *Staphylococcus aureus*, (presumably from a standard collection) the only details given are that the medium is Columbia broth and the pump rate through the calorimeter vessel is 25 cm^3 h^{-1} (Semenitz 1980). The power output is expressed in terms of an electrical output with no indication of a conversion factor. Since other growth parameters were not monitored, detailed interpretation of the *p-t* trace is not possible. This was the control plot for a study of the effect of antibiotics on the growth of this organism; only changes in the shape and position of the *p-t* traces were considered. The lack of information about changes in other growth parameters prevents a quantitative interpretation of the effect of antibacterial agents on the thermal properties of growing cells. Bunker (1985) and Bowden and James (1985c) have shown that not only is there a change in the shape and position of the *p-t* trace during growth in the presence of antibacterial agents but there are significant changes in $\Delta_p H$ and $\Delta_{met} H$. In later papers, Semenitz includes plots of the turbidity of the culture but this is not converted into numbers of cells or biomass. The reporting of power merely as a 'heat response' or as μW without reference to the weight of cells or volume of culture is not acceptable.

1.4 Applications of Calorimetric Studies

Since the previous review (Beezer 1980) there have been many reports of new applications of calorimetric methods and data obtained from them; this is largely the result of a more quantitative approach to thermal measurements with the simultaneous monitoring of other parameters. Applications described and discussed in the following chapters include:

—the evaluation of thermodynamic parameters for microbial growth and metabolism of substrate (p. 75), and the establishment of mass and energy balances for growing organisms;

—a study of the thermodynamics of drug partitioning (p. 167);

—investigations of antimicrobial effects of drugs and antibiotics singly (p. 91) and in combination (p. 93); fungicides (p. 124) and slimicides (p. 100);

—studies of the degradation of solid substrates (p. 99) and the microbial activity of soils and sewage (p. 100);

—establishment of an assay for polyene antibiotics (p. 124);

—studies of the thermal properties of a range of blood cells in anaemia, thyroid diseases and malignancy (p. 131);

—studies of the thermal properties of cells grown in tissue culture, lymphoma cells and mycoplasma and the effects of drugs and metabolic inhibitors on their heat production (p. 147);

—the calculation of oxycaloric equivalents in dissipative metabolism from chemical stoichiometries and the corresponding enthalpies of reaction (p. 192);

—an investigation of the role of endogenous metabolism in ATP synthesis (p. 159);

—ecological studies to provide an insight into how organisms perform and adapt in response to changes in both natural and anthropogenic environmental factors (p. 182).

1.5 Future of Calorimetry in the Study of Cellular Biological Systems

The future for calorimetry in this field is bright, now that the techniques have been established and, with the development of improved instrumentation for monitoring thermal and other supplementary properties of systems, it will be possible to make even more detailed studies of cellular systems. There is a great potential for thermal measurements in basic sciences, medicine, pharmacology and toxicology; for maximum value the results of such measurements must be associated with other chemical and biochemical data for the system under study.

Potential projects, or the extension of current projects, which will benefit from the application of calorimetric techniques include:

—the evaluation of sensitive bioassay techniques for testing the effects of toxicants on the metabolic activity of organisms;

—investigations of the mode of action of, and the design of, drugs;

—a study of synergism and antagonism in drug combinations in a more meaningful way than determining minimum inhibitory concentrations;

—an evaluation of variable complex commercial substrates; including the possibility of microbial growth on hydrocarbons and industrial by-products;

—studies of the variation of interaction of microorganisms with substrates and inhibitors during the cell cycle using synchronous cultures;

—studies involving growth in continuous culture to provide thermal data of relevance in the design of industrial fermentors;

—the construction of energy budgets and the measurement of animals' (e.g. mussel) performance and adaptation in response to environmental change and pollution;

—a rapid evaluation of potential drugs and antibacterial agents, using chemostat cultures;

—the monitoring of changes in soil caused by the addition of fertilizers, biocides, etc.;

—studies of antigen-antibody reactions;

—studies of the relationships between the metabolism of malignant cells and the development of the disease.

References

Bayne-Jones S (1929) *J Bacteriol* **17**, 105–22

Bayne-Jones S and Rhees H S (1929) *J Bacteriol* **17**, 123-40

Beezer A E (1980) *Biological Microcalorimetry*. London, Academic Press

Bowden C P P and James A M (1985) *Microbios* **44**, 201–16

Bunker J C (1985) *The effect of some platinum group metal complexes on bacterial growth*. Ph.D. Thesis, University of London

Dubrunfaut M (1856) *C R Soc Biol* (*Paris*) **42**, 945-8

Hill A V (1912) *J Physiol* **43**, 260–85

Leiseifer H P and Schleser G H (1982) *Z Naturforsch* **38**, 259–67

Nichols S C (1980) *A microcalorimetric study of energy changes during bacterial growth*. Ph.D. Thesis, University of London

Rubner M (1903) *Hyg Rundschau* **13**, 857–64

Rubner M (1908) *Arch Hyg* **66**, 81–92

Semenitz E (1979) *Z Klin Ther Infekt* **7**, 68–70

Shearer C (1921) *J Physiol* **55**, 50–60

Chapter 2

Energy Requirements in Biological Systems

L. E. Erickson

Nomenclature

ATP	adenosine triphosphate
b	mol O_2 per quantity of organic substrate containing 1 g atom carbon; see Equation (2.1)
d	mol CO_2 per quantity of organic substrate containing 1 g atom carbon; see Equation (2.1)
g	free energy, kJ (equivalent of available electrons) $^{-1}$
ΔH_f	heat of fermentation, kJ (quantity of substrate containing 1 g atom carbon) $^{-1}$
m_{ATP}	maintenance coefficient, g mol ATP (g biomass) $^{-1}$ h $^{-1}$
m_e	maintenance coefficient, equivalent of available electrons per equivalent of available electrons h $^{-1}$
m_{th}	maintenance coefficient in free energy units, h $^{-1}$
NAD	oxidized and reduced (NADH) forms of nicotinamide adenine dinucleotide
NADP	oxidized and reduced (NADPH) forms of nicotinamide adenine dinucleotide phosphate
Q	enthalpy, kJ (eq available electrons) $^{-1}$
Y_{ATP}	biomass yield, g biomass (mol ATP) $^{-1}$
Y_{ATP}^{max}	true growth yield, g biomass (mol ATP) $^{-1}$
$Y_{p/ATP}^{max}$	true product yield, g product (mol ATP) $^{-1}$
y_c	biomass carbon yield; see Equation (2.1)
z	Product carbon yield; see Equation (2.1)

Greek

γ	reductance degree, eq of available electrons per g atom carbon
δ	equivalents of available electrons transferred to produce 1 mol of ATP

ϵ	fraction of available electrons transferred to oxygen, $\epsilon = 4b/\gamma_s$
η	biomass yield in available electron units
η_e	biomass energetic yield
η_{max}	true growth yield in available electron units
η_{th}	biomass free energy yield
η_{th}^{max}	true growth yield in free energy units
μ	specific growth rate, h^{-1}
σ	weight fraction carbon

Subscripts

b	biomass
p	product
s	substrate

2.1 Introduction

Most biological systems are open systems with flow of mass to and from the system. Primary producers convert solar radiation to chemical energy. Carbohydrates are produced from carbon dioxide, water, and solar energy by plants and photosynthetic microorganisms. The carbohydrates are the starting point for many biological processes which occur because the free energy of the carbohydrates is sufficient to provide for the energy requirements of the biological system. These biological processes obey the laws of thermodynamics and the atomic species conservation laws.

The literature on bioenergetics is very extensive and covers a wide variety of topics. In this work the emphasis will be directed toward cellular biological systems in which chemical energy provides for the energetic needs of the system. Most of the attention will be on single cell systems; however, the basic conservation equations are applicable to other systems also. Bioenergetics and cellular metabolism are covered in biochemistry texts (Lehninger 1982); thermodynamic considerations are covered in biochemical engineering books (Roels 1983). No attempt will be made to exhaustively cover these topics in this work.

2.2 Theory

The total mass balance, carbon balance, available electron balance, and nitrogen balance are the mass balances that are commonly used in the analysis of biological systems (Erickson *et al* 1978a; Erickson 1979; Erickson and Patel 1982a; Roels 1983). An energy balance can also be written to describe the flow of energy associated with the process. The energy balance is important in calorimetric studies as the flow of heat to or from the system is one of the terms in the energy balance. The free energy requirements, associated with biological transformations and the thermodynamic efficiency of biological systems, depend on the free energy of the reactants and products. The free energy (i.e. Gibbs function) of a compound depends on concentration as well as temperature, while the enthalpy can be assumed to depend only on temperature. In calorimetric studies (Beezer 1980), the enthalpies of the reactants and products are needed in the energy balance; each enthalpy is frequently assumed to depend only on temperature when estimates are made for this purpose.

2.2.1 Energy Regularities

There are several regularities (Minkevich and Eroshin 1973) which are important to consider in discussing the energy requirements in biological systems. An important energy regularity was reported by Thornton (1917). As shown in Table 2.1, Thornton reported that the heat of combustion of a wide variety of organic compounds is relatively constant when the enthalpy change per equivalent of oxygen utilized is measured. Patel and Erickson (1981) have reviewed the work of Thornton and later workers; they have shown that this regularity is very useful in the analysis of biological systems (Erickson and Patel 1982a). One of the oldest and most widely used examples of this regularity is the process of measuring oxygen consumption to estimate metabolic heat production in medical work (Kleiber 1961; Seagrave 1971). This regularity has also been used with measurements of heat evolution from a biological reactor in the analysis of fermentation processes (Cooney *et al* 1969; Erickson *et al* 1978b; Luong and Volesky 1980; Volesky *et al* 1981). Minkevich and Eroshin (1973) played an important role in identifying the importance of this regularity in the analysis of microbiological processes.

Erickson and Patel (1982a) and Roels (1983) have examined the close relationship between the free energy of combustion and the heat of combustion of organic compounds with special attention directed toward those compounds of biological importance. Erickson and Patel (1982a) have shown that the free energy of combustion is approximately equal to the heat of

Table 2.1 Regularity values for heat evolved per equivalent of oxygen consumed, Q

Substrate	Q/kJ (eq of available electrons)$^{-1}$	*Reference*
organic compounds	111	Thornton 1917
fats	110	Kleiber 1961
proteins	106	Kleiber 1961
carbohydrates	117	Kleiber 1961
microbial biomass	113	Minkevich and Eroshin 1973
microbial biomass	117	Roels 1983
glucose	116	Volesky *et al* 1981
monosaccharides	117	Erickson and Patel 1982a
proteins	109	Erickson and Patel 1982a
fats	109	Erickson and Patel 1982a

combustion. Roels (1983) has found that the free energy of combustion per equivalent of oxygen consumed depends to a small extent on the reductance degree of the compound being oxidized. Methane, which has a reductance degree of 8 equivalents per mole of carbon, has one of the smallest free energy changes per equivalent (102·3 kJ eq^{-1}). Glucose (119·7 kJ eq^{-1}) and other sugars have relatively large changes in free energy per equivalent (for other tabulated values, see Patel 1981; Erickson and Patel 1982b; and Roels 1983 pp 40, 41). These small differences in free energy level are very important in anaerobic fermentations because these differences are the only driving force which causes the anaerobic process to proceed in fermentations where all of the available electrons in the substrate are converted to microbial biomass and products such as ethanol, butanol, and lactic acid. In anaerobic digestion, a wide variety of organic compounds can be converted to methane and carbon dioxide because the free energy level of methane per equivalent is less than the regularity value of 108·7 kJ eq^{-1} (Erickson and Patel 1982a). All large biological organic compounds appear to have free energy levels which fall somewhere between the higher values for carbohydrates (120 kJ eq^{-1}) and the low value for methane.

The energy requirements in biological systems are easily met because of the energetic regularities. Carbohydrates are widely used as a source of chemical energy for cellular processes. The ability of a microbial cell to grow on glucose and mineral salts is possible because the energy level of all the organic compounds that are important in cellular biochemistry is relatively similar when viewed in terms of free energy per equivalent of available electrons. The energy required to rearrange available electrons to form the biological monomers and polymers is not large. The high energy bonds associated with adenosine triphosphate (ATP) are sufficient to provide the energy needed for a wide variety of biochemical reactions.

The free energy and enthalpy per equivalent of available electrons may be considered relative to the combustion dead state of liquid water and the gases carbon dioxide, nitrogen, and sulphur dioxide at 25 °C and one atmosphere pressure. In biological systems, the physiological dead state of liquid water, liquid urea, carbon dioxide gas, and hydrogen sulphide gas at one atmosphere and 25 °C may be used. When examining tabulated values of free energy and enthalpy, the dead state used in the table should be carefully noted. Values have been tabulated by Minkevich (1980), Patel (1981), Patel and Erickson (1981), Erickson and Patel (1982a, b) and Roels (1983) using units which present the values on a per equivalent of available electrons basis.

2.2.2 Balances and Yields

Consider the balance equation for heterotrophic growth and product formation:

$$CH_mO_k + aNH_3 + bO_2 = y_cCH_pO_nN_q + zCH_rO_sN_t + cH_2O + dCO_2 \qquad (2.1)$$

where CH_mO_k denotes the elemental composition of the organic substrate, $CH_pO_nN_q$ is the elemental composition of the biomass, and $CH_rO_sN_t$ is the elemental composition of the extracellular product. The carbon balance is:

$$y_c + z + d = 1{\cdot}0 \tag{2.2}$$

where y_c, z, and d give the fraction of organic substrate carbon converted to biomass, extracellular product, and CO_2, respectively. The available electron balance may be written in the form:

$$\frac{4b}{\gamma_s} + y_c\frac{\gamma_b}{\gamma_s} + z\frac{\gamma_p}{\gamma_s} = 1{\cdot}0 \tag{2.3}$$

provided the physiological dead state is used to define the reductance degree, γ. When the valences $C = 4$, $H = 1$, $O = -2$, and $N = -3$ are used, NH_3, CO_2, and H_2O have zero available electrons and the equations:

$$\gamma_s = 4 + m - 2k \tag{2.4}$$

$$\gamma_b = 4 + p - 2n - 3q \tag{2.5}$$

$$\gamma_p = 4 + r - 2s - 3t \tag{2.6}$$

give the reductance degree of the substrate, biomass, and product, respectively, based on their elemental composition. The value of $\gamma_b = 4{\cdot}291$ has been used as an average value for Equation (2.5) when biomass elemental composition is not available (Erickson 1980a).

The available electron balance may be written in the form:

$$\epsilon + \eta + \xi_p = 1{\cdot}0 \tag{2.7}$$

by defining $\epsilon = 4b/\gamma_s$ as the fraction of available electrons in the organic substrate which are transferred to oxygen, $\eta = y_c\gamma_b/\gamma_s$ as the biomass yield in available electron units, and $\xi_p = z\gamma_p/\gamma_s$ as the product yield in available electron units.

The enthalpy and free energy are quantities which are defined relative to a selected dead state or zero energy state. If the physiological dead state is defined such that the ammonia, water, and carbon dioxide have zero energy level, then energetic yields analogous to the available electron yields can be written for enthalpy:

$$\eta_e = \frac{y_c\gamma_b Q_b}{\gamma_s Q_s} \tag{2.8}$$

$$\xi_{pe} = \frac{z\gamma_p Q_p}{\gamma_s Q_s} \tag{2.9}$$

where Q refers to the enthalpy per equivalent of available electrons. For free energy:

$$\eta_{th} = \frac{y_c \gamma_b g_b}{\gamma_s g_s} \tag{2.10}$$

$$\xi_{th} = \frac{z \gamma_p g_p}{\gamma_s g_s} \tag{2.11}$$

where g refers to the free energy per equivalent of available electrons relative to the selected dead state. If the regularity value is used and $Q_b = Q_p = Q_s = g_b = g_s = g_p = Q_0$ is assumed, then the energetic yields are equal to the available electron yields. For this case, the heat of fermentation associated with Equation (2.1) is directly related to ϵ; that is:

$$\Delta H_F = 4bQ_0 \tag{2.12}$$

where ΔH_F is the heat evolved per mol of carbon substrate consumed and Equation (2.12) makes use of Thornton's rule (Patel and Erickson 1981). Some representative values of Q_0 are given in Table 2.1.

In the case of aerobic cultures, use of the regularity value Q_0 is appropriate because of the current level of accuracy with which fermentation measurements are made. The variation in the reported values of Q_0 due to experimental error appears to be relatively large in several cases. Cooney *et al* (1969) report 130 kJ per equivalent of available electrons. Luong and Volesky (1980) report values ranging from 86 to 150 kJ eq^{-1}, with an average value of 114 kJ eq^{-1}. Erickson *et al* (1978b) report values ranging from 50·2 to 172 kJ eq^{-1} with an average value of 111 kJ eq^{-1}. The largest errors occur where the rates are small and measurement errors are expected to be larger on a percentage basis. Erickson (1980a) has attributed much of the variation in values of Q_0 to experimental error; however, values for carbohydrates are known to be larger than the values for proteins and fats.

For the case where conditions are anaerobic, b = 0 in Equation (2.1). When there are no external electron acceptors:

$$\eta + \xi_p = 1{\cdot}0 \tag{2.13}$$

and the heat of fermentation is:

$$\Delta H_F = Q_s \gamma_s \left(1 - \frac{Q_b}{Q_s}\eta - \frac{Q_p}{Q_s}\xi_p\right) \tag{2.14}$$

The average value of 113 kJ eq^{-1} may be used for microbial biomass in

Equation (2.14) when values of Q_b are not determined experimentally (Minkevich and Eroshin 1973; Erickson 1980a).

In anaerobic fermentations, electron acceptors such as nitrate and sulphate can provide conditions in which ATP is formed as energy is dissipated by transferring electrons from the energy source (the organic substrate) to the electron acceptor. Roels (1983) has tabulated the free energy change per equivalent of available electrons for a variety of electron acceptors. When nitrate is converted to nitrogen gas according to the reaction:

$$2NO_3^- + 7H^+ + 5NADH = N_2 + 6H_2O + 5NAD^+ \tag{2.15}$$

The free energy change is 111·91 kJ eq^{-1} of available electrons. Thus, the bioenergetics when nitrate is the electron acceptor are relatively close to those with oxygen as the electron acceptor. If nitrate is used as the nitrogen source for growth or microbial product formation, energy is required for nitrogen reduction to the amino form ($N = -3$ valence). Because of this, growth yields in mass units are generally smaller when nitrate is used to replace ammonia as the nitrogen source (Roels 1983).

2.3 Efficiency of Biological Systems

The efficiency with which biological systems function has received much attention. The efficiency of growth and product formation is important in conventional agriculture, microbial processes, and in other applications. Erickson and Patel (1982a) have reviewed process efficiencies in living systems. The utilization of energy for growth, product formation and maintenance has been considered for microbial cultures, milk cows, and hens (Brody 1945; Erickson and Patel 1982a; Roels 1983). The efficiency with which free energy in food is converted to mechanical work by living systems has also been investigated (Brody 1945; Ricci 1967; Astrand and Rodahl 1977; Erickson and Patel 1982a). The conversion of chemical energy to mechanical work will not be considered further here.

When the efficiency of growth and product formation is considered, experimental values may be compared to the maximum thermodynamic efficiency and to the maximum biochemical efficiency based on knowledge of the biochemical pathways and stoichiometry of individual reactions and transport processes. The maximum thermodynamic efficiency is 100%. Stouthamer (1979) has reviewed some of the biochemical pathway energetic requirements by estimating the moles of ATP required for growth on various media. Roels (1983) has reviewed actual growth yields and compared them to maximum yields. While some information is available on product yields, little has been done to systematically look at biochemical limitations for product formation efficiency. Product yields corrected for growth and maintenance have been estimated in only a limited number of cases (Erickson 1980b, c; Erickson and Patel 1982a; Oner *et al* 1983).

The utilization of organic substrate for growth and maintenance by microbial cultures has been extensively treated by Pirt (1965, 1975), Stouthamer (1979), Roels (1983), and Solomon *et al* (1984). Using available electron units, one may write:

$$\frac{\mu}{\eta} = \frac{\mu}{\eta_{max}} + m_e \tag{2.16}$$

where η_{max} is the growth yield corrected for maintenance or true growth yield and m_e is the maintenance coefficient. Roels (1983) has pointed out that organic substrate limited heterotrophic growth may be energetically-limited or carbon-limited. Figures 2.1 and 2.2 show how the carbon yield, y_c and the available electron yield, η, vary with the reductance degree, γ_s, of the organic substrate. For values of $\gamma_s > 4{\cdot}67$, the results suggest that growth may be carbon-limited rather than energy-limited (Erickson 1981). Since most studies have been carried out with one or more carbohydrates as the organic substrate, substrate-limited growth is frequently energy-limited growth.

The true growth yield and the maintenance coefficient have been estimated by a number of investigators (Pirt 1975; Roels 1983; Solomon *et al* 1984; Lee

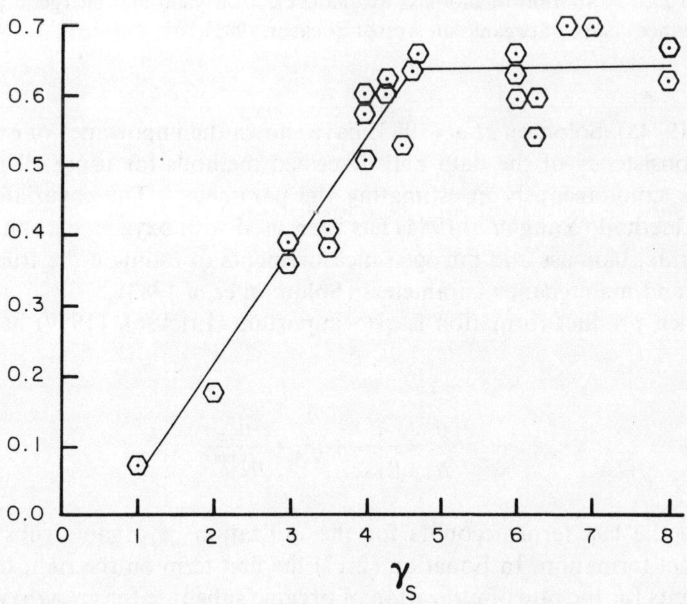

Figure 2.1. Variation of biomass carbon yield with reductance degree of organic substrate (Erickson 1981).

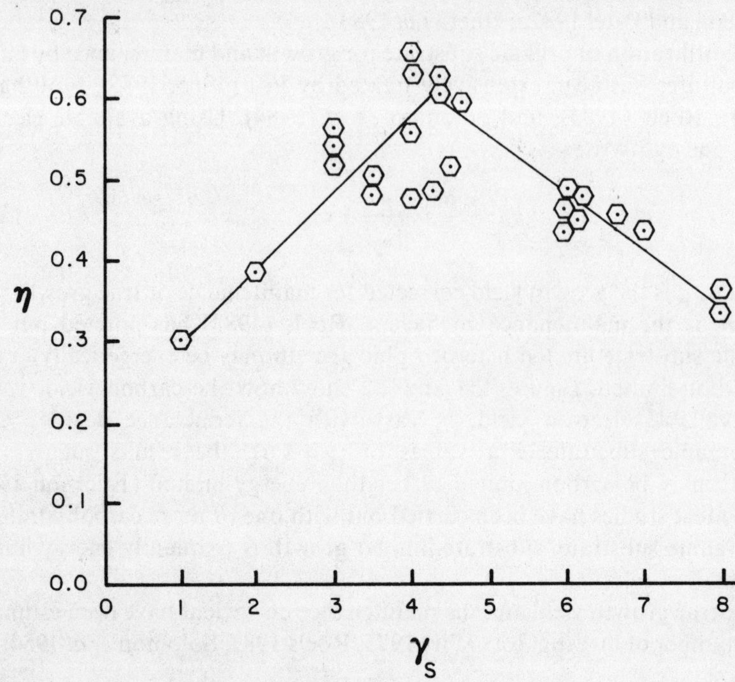

Figure 2.2. Variation of biomass available electron yield and energetic yield with reductance degree of organic substrate (Erickson 1981).

et al 1984a). Solomon *et al* (1983) have shown the importance of examining the consistency of the data and described methods for using all measurements simultaneously in estimating the parameters. The covariate adjustment method (Yang *et al* 1984) has been used with oxygen, carbon dioxide, substrate, biomass and nitrogen measurements to estimate the true growth yield and maintenance parameters (Solomon *et al* 1983).

When product formation is also important, Erickson (1979) has shown that:

$$\frac{\mu}{\eta} = \frac{\mu}{\eta_{\max}} + m_e + \frac{\mu \xi_p}{\eta \xi_p^{\max}} \tag{2.17}$$

where the last term accounts for the utilization of organic substrate for product formation. In Equation (2.17) the first term on the right hand side accounts for the rate of utilization of organic substrate for growth, with $\eta_{\max}$ being the true growth yield or growth yield corrected for maintenance and product formation. The second term on the right hand side is the rate of utilization of organic substrate for maintenance; the third term is the rate of utilization of organic substrate for product formation. In Equation (2.17), ξ_p^{max} is the true product yield or the product yield corrected for growth and maintenance.

The energy requirements for growth and product formation are generally provided by organic molecules which contain the available electrons which are used to form the biomass and product. Most of the ATP which is used in biosynthesis to rearrange the available electrons of the substrate into cellular monomers and biopolymers is provided by oxidative phosphorylation, substrate phosphorylation, or photophosphorylation. The formation of ATP from ADP by oxidative phosphorylation requires an electron acceptor such as oxygen, nitrate, nitrite, or sulphate. Substrate phosphorylation requires conversion of the substrate to a product with a lower free energy level such as the conversion of glucose to lactic acid. Photophosphorylation utilizes light energy to provide the increased energy associated with ATP. In photosynthesis, the available electrons needed for biosynthesis can be obtained by reduction of organic molecules such as NADP using light energy (Lehninger 1982). The conversion of radiant energy to chemical free energy is limited by the Carnot efficiency (Erickson and Patel 1982a). Because of this, the actual thermodynamic efficiencies associated with photosynthetic growth are quite low.

The available electron yield, η, and the yield based on ATP, Y_{ATP}, are related according to the following expression:

$$\eta = \frac{\begin{array}{c}\text{available electrons}\\ \text{incorporated into biomass}\end{array}}{\begin{array}{c}\text{available electrons}\\ \text{incorporated into biomass}\end{array} + \begin{array}{c}\text{available electrons}\\ \text{used to generate ATP}\end{array}}$$

$$\eta = \frac{\sigma_b \gamma_b Y_{ATP}/12}{\dfrac{\sigma_b \gamma_b Y_{ATP}}{12} + \delta} \tag{2.18}$$

where Y_{ATP} has units of g biomass (mol ATP)$^{-1}$ and δ is equivalents of available electrons transferred to generate one mol of ATP from ADP and inorganic phosphate. In aerobic fermentations with glucose as substrate, 38 mol of ATP may be generated per mol of glucose oxidized to CO_2 and H_2O. In this case $\delta = 24/38 = 0{\cdot}632$ and the available electrons used to generate ATP are transferred to oxygen. For glucose as substrate with anaerobic production of ethanol, if two mol of ATP are formed per mol of glucose, $\delta = 24/2 = 12$ equivalents of available electrons transferred to ethanol per mol ATP.

When the organic substrate is used for growth, maintenance, and product formation, each may be separately related to ATP as follows:

$$\eta_{\max} = \frac{\sigma_b \gamma_b Y_{\mathrm{ATP}}^{\max}/12}{\dfrac{\sigma_b \gamma_b Y_{\mathrm{ATP}}^{\max}}{12} + \delta} \tag{2.19}$$

$$m_e = \frac{12\delta m_{\mathrm{ATP}}}{\sigma_b \gamma_b} \tag{2.20}$$

$$\xi_p^{\max} = \frac{\sigma_p \gamma_p Y_{\mathrm{p/ATP}}^{\max}/12}{\dfrac{\sigma_p \gamma_p Y_{\mathrm{p/ATP}}^{\max}}{12} + \delta} \tag{2.21}$$

The average value of $\sigma_b = 0{\cdot}462$ may be used for the weight fraction carbon in dry biomass when the exact elemental composition is unknown (Minkevich and Eroshin 1973; Erickson 1980a).

These equations allow the available electron yield and maintenance parameters to be related to parameters that have been extensively investigated. The parameter δ is related to the efficiency with which ATP is formed, while the other parameters, Y_{ATP}, $Y_{\mathrm{ATP}}^{\max}$, m_{ATP} and $Y_{\mathrm{p/ATP}}^{\max}$ are related to the efficiency with which ATP is utilized. Theoretical values of $Y_{\mathrm{ATP}}^{\max}$ have been reported by Stouthamer (1973; 1977a, b; 1979), van Dijken and Harder (1975), and Harder and van Dijken (1976). Some of these results are presented in Table 2.2. For organic substrates such as methane, methanol, and n-hexadecane, the potential for carbon limitation is relatively great. Erickson (1981) has shown that carbon limitation influences the theoretical yield based on biochemical pathway information when n-hexadecane is the organic substrate. For organic substrates such as n-alkanes, methane, and methanol which must be oxidized to obtain the monomers and polymers needed in biomass, the initial oxidation may occur without recovery of any energy in the form of ATP. For these substrates it is difficult to use Equation (2.18) directly because the ATP production associated with several different oxidations must be considered; if there is oxidation without formation of ATP, this also must be included in the denominator of Equation (2.18).

Yield data for microorganisms and other biological systems has been widely reported in the literature. Energetic yields have been reported by Minkevich and Eroshin (1973), Erickson (1981), Erickson and Patel (1982a),

Table 2.2 Estimated maximum values of Y_{ATP}^{max}, based on biochemical pathway analysis

Organic substrate	*Nitrogen source*	Y_{ATP}^{max}/g biomass (mol ATP)$^{-1}$	*Remarks*
glucose	NH_3	28·8	
glucose, amino acids and nucleic acid bases	NH_3	31·9	
glucose	NO_3^-	23·1	
glucose	N_2	11·1	12 mol ATP required to fix 1 mol N_2
pyruvate	NH_3	13·5	
lactate	NH_3	13·4	
malate	NH_3	15·4	
acetate	NH_3	10·0	
CO_2 + NADPH	NH_3	6·5	Calvin cycle
methanol	NH_3	28·4	ribulose phosphate pathway of formaldehyde fixation
methane	NH_3	12·5	serine pathway

Roels (1983), and many others. Payne (1970) has tabulated microbial yields in several different sets of units.

For carbohydrates such as glucose and other simple sugars as substrates, microbial growth yields have an average value of $\eta = 0{\cdot}59$ for aerobic growth with ammonia as nitrogen source (Solomon and Erickson 1981). This value is also the energetic yield, η_e, based on the physiological dead state of CO_2, H_2O and ammonia. Roels (1983) reports a thermodynamic efficiency of 0·64 based on the combustion dead state for glucose. These values are about 67–70% of the corresponding maximum values based on an analysis of biological pathways. For example, the estimated value from Equation (2.18) for $Y_{ATP}^{max} = 28{\cdot}8$ and $\delta = 0{\cdot}632$ is $\eta_{max} = 0{\cdot}88$; thus for the physiological dead state, $0{\cdot}59/0{\cdot}88 = 0{\cdot}67$ or 67%. Erickson (1981) has reported that growth yields on hydrocarbons are also about 67% of the maximum value estimated by pathway analysis ($0{\cdot}41/0{\cdot}61 = 0{\cdot}67$). Some of the available yield data in the literature for aerobic growth processes is presented in Figures 2.1 and 2.2 and Tables 2.3, 2.4, and 2.5. Table 2.3 presents some results in which aerobic microbial growth is compared to growth of an embryo for silk worms, chickens, sea urchins, and frogs. When the energetic growth requirements of rats, pigs, lambs, and cattle are corrected for maintenance, values ranging from 0·59 to 0·78 are obtained for the true growth yield, η_{max} (Erickson and

Table 2.3 Comparison of biomass energetic yield for microbial growth on glucose with prenatal growth[a]

Type of biological growth	η
microbial growth	0·59
silk worm embryo	0·65
chick embryo	0·63
sea urchin embryo	0·59
frog embryo	0·51

[a]Brody (1945), Solomon and Erickson (1981) and Erickson and Patel (1982a).

Patel 1982a). Thus, the differences in growth yield are relatively small when one considers the wide range of growth processes which are compared.

The energetic requirements for growth in terms of the fraction of energy incorporated into microbial biomass are relatively similar for many different microorganisms growing on the same organic substrate (Solomon and Erickson 1981; Roels 1983). There are significant differences in growth yields when different organic substrates are used as shown in Figures 2.1 and 2.2 and Table 2.4. The highest yields in Table 2.4 are for glucose and glycerol. The results in

Table 2.4 Thermodynamic efficiency of aerobic microbial growth based on the combustion dead state; results of Roels (1983)

Substrate	γ_s	η_{th}
methane	8	0·31
n-alkanes	6·1	0·42
methanol	6	0·40
ethanol	6	0·40
glycerol	4·7	0·62
mannitol	4·3	0·55
acetate	4	0·48
lactate	4	0·57
glucose	4	0·64
formaldehyde	4	0·48
gluconate	3·7	0·59
succinate	3·5	0·51
citrate	3·5	0·525
malate	3	0·535
formate	2	0·46
oxalate	1	0·23

Table 2.5 True growth yields and true product yields for milk production by cattle and polysaccharide production by bacteria

Type of biological growth	η_{max}	ξ_p^{max}	*Maintenance coefficient*
cattle	0·71	0·74 (milk)	598·3 kJ $kg^{-0.75}$ day^{-1}
cattle	0·78	0·76 (milk)	608·2 kJ $kg^{-0.75}$ day^{-1}
cattle	0·65	0·66 (milk)	547·4 kJ $kg^{-0.75}$ day^{-1}
bacteria	0·66	0·89 (polysaccharides)	0·004 h^{-1}
bacteria	0·63	0·78 (polysaccharides)	0·007 h^{-1}

Table 2.4 and Figure 2.2 have similar trends. Growth yields are smallest for values of γ_s far from γ_b; this result is logical in that molecules such as oxalic acid and methane must be modified substantially before they can be incorporated into the main components of microbial biomass.

Table 2.5 presents some true growth yields and true product yields for microbial polysaccharide production and milk production from cattle. For milk, the true product yields are similar in magnitude to the true growth yields. For polysaccharides, the true product yields are somewhat higher; this is expected since the process of polysaccharide formation has a relatively large theoretical true product yield of 0·95 (Erickson 1980a) based on an analysis of the biochemical pathway.

For anaerobic growth, the free energy yield or thermodynamic efficiency has been considered by Roels (1983) for the combustion dead state and by Erickson and Oner (1983), and Oner *et al* (1984), Oner (1985) for the physiological dead state. Table 2.6 presents the growth yield in free energy units based on results reported by Roels (1983). Roels (1983) shows that the effect of culture conditions on the value of the free energy of combustion may affect

Table 2.6 Thermodynamic efficiency of anaerobic microbial growth without external electron acceptors; results of Roels (1983) for combustion dead state

Organic substrate	*Products*	η_{th}
glucose	ethanol	0·67
glucose	lactate	0·69–0·83
glucose	propionate and acetate	0·79
glucose	acetate and methane	0·77
acetate	methane	0·46–0·70
methanol	methane	0·63
formate	methane	0·24–0·43
propionate	methane	0·58–0·73

Table 2.7 Values of true free energy biomass yield, η_{th}^{max}, based on physiological dead state for anaerobic fermentations without external electron acceptors; results of Oner (1985) and Oner *et al* (1984)

Organic substrate	*Products*	η_{th}^{max}	*Remarks*
glucose	lactate	0·87	Data of Luedeking and Piret (1959)
glucose	lactate	0·89	Data of Hanson and Tsao (1972)
crude sorghum extract and vetch juice	lactate	0·54–0·73	Data of Samuel *et al* (1980)
glucose	lactate, acetate, ethanol, formate	0·82	Data of DeVries *et al* (1970)
glucose	acetone, butanol ethanol, butyrate	0·74	Data of Leung and Wang (1981)
methanol	butyrate and acetate	0·84–0·89	Data of Datta and Ogeltree (1983)
lactose (milk)	lactate	0·64–0·80	Data of Oner (1985)

the results. In Table 2.6 the range of values reported for the substrates acetate, formate, and propionate are due to free energy differences between standard state conditions and pH 7. Because of the small differences in free energy that power anaerobic growth, the estimation of the free energy growth yield and thermodynamic efficiency is not as precise for anaerobic processes as it is for aerobic processes. Some additional results for anaerobic processes from the work of Oner (1985) are presented in Table 2.7. Table 2.8 presents some calculated values of the available electron yield, η_{max}, based on Equation

Table 2.8 Comparison of true growth yields for anaerobic ethanol and lactate fermentations with glucose as organic substrate for given values of δ and Y_{ATP}^{max}

Product	δ	Y_{ATP}^{max}	η_{max}	η_{th}^{max}
ethanol	12	10·5	0·126	0·611
		28·8	0·284	0·795
		31·9	0·305	0·808
ethanol	24	10·5	0·0674	0·450
		28·8	0·165	0·678
		31·9	0·180	0·698
lactate	12	10·5	0·126	0·744
		28·8	0·284	0·868
		31·9	0·305	0·876

(2.19) and the corresponding free energy yield:

$$\eta_{th}^{max} = \frac{g_b \eta_{max}}{g_s - g_p(1 - \eta_{max})} \tag{2.22}$$

for the values of δ and Y_{ATP}^{max} indicated in the table. Comparison of Table 2.7 with Table 2.8 shows that some of the estimated values of true growth yield are larger than the predicted maximum values based on pathway analysis. As discussed elsewhere (Oner *et al* 1984), the reasons for this are not known; the presence of electron acceptors could account for the increased yields which led to these results.

Examination of Tables 2.4, 2.6, and 2.7 reveals that the free energy biomass yield is much higher for methanol under anaerobic conditions than it is under aerobic conditions. Several of the other free energy yields are also higher under anaerobic conditions compared to aerobic conditions.

Growth yields for photoautotrophic growth and mixotrophic growth have been reported recently by Lee *et al* (1984b, 1985). Equations and methods for estimating energetic yields as well as a summary of available results are presented in their work. Energy requirements for these systems depend on how the available electrons for biomass synthesis are supplied. Where solar energy is used to reduce NADP, with the hydrogen supplied by water, radiant energy is required to power this process (Lehninger 1982). The thermodynamic efficiency with which radiant energy is converted to chemical energy in the form of NADPH must be less than the Carnot efficiency. Additional energy is required to convert CO_2 and NADPH into microbial biomass. For $Y_{ATP}^{max} = 6{\cdot}5$, and $\delta = 0{\cdot}667$, $\eta_{max} = 0{\cdot}62$; that is, the energetic yield of converting CO_2 and NADPH to microbial biomass is about 0·62 or less. When both the light and dark processes are considered together, the maximum estimated energetic yield is about 0·29 (Lee *et al* 1984b). Most reported yields are less than 0·29; however, some experimental results exceed this predicted maximum yield (Lee *et al* 1984b).

2.4 Maintenance Energy Requirements

In the analysis of substrate utilization for growth, maintenance, and product formation, maintenance coefficients are estimated. Maintenance coefficients may be expressed in several different sets of units. Equations (2.16), (2.17) and (2.20) consider m_e with units of equivalents of available electrons utilized for maintenance per equivalent of available electrons of biomass per hour. For aerobic processes, utilization of the energy regularities allows m_e to also be considered with units of kJ of energy expended for maintenance per kJ of biomass energy per hour. For anaerobic processes, the maintenance coefficient in energetic units is:

$$m_{th} = \frac{(g_s - g_p)\, m_e}{g_b} \tag{2.23}$$

Roels (1983) reviews values of the maintenance coefficient for a wide variety of aerobic growth processes and reports an average value of 0·3 eq of available electrons per mol of carbon in microbial biomass. Using the physiological dead state, this corresponds to about 0·07 h^{-1}; this is about 35 kJ per mol of biomass carbon per hour. Roels (1983) also reports that the maintenance energy requirements may increase with temperature; the activation energy reported is 40 kJ mol^{-1}. Stouthamer (1979) reports values of m_{ATP} for anaerobic growth which range from 0·2 to 38·7 mmol of ATP (g cell) $^{-1}$ h^{-1}. Energetically, this corresponds approximately to values of m_{th} ranging from 0·0004 to 0·08 h^{-1}. Oner *et al* (1984) have recently reviewed some of the available data in the literature for anaerobic growth. Values of the maintenance coefficient range from values that are so small that they are not significantly different from zero to relatively large values for organisms such as *Zymomonas mobilis* where m_{th} ranges from 0·15 to 0·47 h^{-1}. In recent experimental work with yoghurt cultures with dry milk as substrate, maintenance coefficients ranged from about 0·003 to 0·03 h^{-1} (Oner 1985). Pirt (1975) has pointed out that some of the variation in maintenance energy requirements is due to osmotic work to maintain concentration gradients between the cell and extracellular media. Increasing ionic concentrations have been observed to result in larger values of the maintenance coefficient. Larger maintenance coefficients have also been observed when the growth process is not energy limited.

2.5 Heat of Fermentation

The energy that is dissipated as heat by the microorganisms or by larger organisms is directly related to the oxygen utilized by the organism under aerobic conditions. Equation (2.12) has been used for aerobic microbial growth, in animal and human metabolic studies, and elsewhere.

Under anaerobic conditions, Equations (2.13) and (2.14) may be used to estimate the heat of fermentation when no external electron acceptors are present in the culture. When electron acceptors such as nitrate, nitrite, and sulphate are used, an enthalpy balance should be made considering all of the reactants and products (Roels 1983). The enthalpy of the reactants relative to the dead state minus the enthalpy of the products (including the biomass) relative to the dead state, gives the heat of fermentation.

The heat of fermentation results from the expenditure of energy for transport, osmotic work, chemical transformations, movement of the microorganism, maintenance of electrochemical potentials, and other cellular activities. Since processes are not 100% efficient some heat is generated in conjunction with many different cellular processes. Fortunately, thermodynamics allows us to use macroscopic mass and energy balances to obtain the heat of fermentation without determining the energetics of each indivdual cellular activity.

The heat of fermentation may leave the fermentor in several different ways. The terms which should be included in an energy balance depend on the design and operation of the system. Examples of energy balances are given by Cooney (1969), Beezer (1980), Luong and Volesky (1980) and Roels (1983). Mechanical energy associated with agitation, evaporative cooling, and heat transfer through the walls of small laboratory fermentors can be important. The enthalpy associated with streams entering and leaving the fermentor should be considered in the energy balance; cooling water, gas, and media flow rates and temperatures should be measured. Stream composition measurements are also needed. The consistency of the results can be checked by relating the mass and energy balance results to the regularities and known thermochemical properties.

2.6 Summary

There are several factors which influence the energy requirements of biological processes and the observed energetic yield. The nature of the source or sources of carbon and energy are important. In a complex medium, some monomers do not need to be synthesized. The ATP production by substrate phosphorylation depends on the pathway of substrate breakdown. The nature of the nitrogen source can influence the bioenergetics in several ways. Energy is required to convert NO_3^- or N_2 to amino nitrogen. The electron acceptor is important; oxygen is the best electron acceptor for most applications. Under anaerobic conditions, other electron acceptors can help provide for cellular energetic requirements by participating in oxidation-reduction reactions that allow the cell to produce ATP. The efficiency of biological processes depends on the efficiency with which energy is incorporated into ATP high energy bonds as well as the efficiency with which ATP is utilized for growth and product formation. The energy regularity allows growth and product formation to be visualized as processes in which molecules of very similar energy level react to form other molecules with similar energy levels. The high energy bonds associated with ATP release more than enough free energy to compensate for the small differences in the free energy per equivalent of biologically useful organic molecules.

Acknowledgement

This work was partially supported by the National Science Foundation.

References

Anstrand P O and Rodahl K (1977) *Textbook of Work Physiology*. New York, McGraw-Hill, pp 1–669

Beezer A E (1980) *Biological Microcalorimetry*. New York, Academic Press, pp 1–483

Brody S (1945) *Bioenergetics and Growth*, New York, Reinhold, pp 1–1023

Cooney C L, Wang D I C and Mateles R I (1969) *Biotechnol Bioeng* **11**, 269–81

Datta R and Ogeltree J (1983) *Biotechnol Bioeng* **25**, 991–8

DeVries W, Kapteijn W M C, Van Der Beek E G and Stouthamer A H (1970) *J Gen Microbiol* **63**, 333–45

Erickson L E (1979) *Biotechnol Bioeng* **21**, 725–43

Erickson L E (1980a) *Biotechnol Bioeng* **22**, 451–6

Erickson L E (1980b) *J Fermentation Technol* **58**, 53–9

Erickson L E (1980c) *Biotechnol Bioeng* **22**, 1929–44

Erickson L E (1981) *Biotechnol Bioeng* **23**, 793–803

Erickson L E and Oner M D (1983) *Ann NY Acad Sci* **413**, 99–113

Erickson L E and Patel S A (1982a) *CRC Crit Rev Biomed Eng* **8**, 311–33

Erickson L E and Patel S A (1982b) *Development of Regularities and Characterization of Renewable Energy Resources*. Alternate Energy Sources **1**, No 4. New York, Comtex Scientific, pp 1–72

Erickson L E, Minkevich I G and Eroshin U K (1978a) *Biotechnol Bioeng* **20**, 1595–621

Erickson L E, Selga S E and Viesturs U E (1978b) *Biotechnol Bioeng* **20**, 1623–38

Hanson T P and Tsao G T (1972) *Biotechnol Bioeng* **14**, 233–52

Harder W and Van Dijken J P (1976) In: Schlegel H P *et al* (eds) *Microbial Production and Utilization of Gases*. Gottingen, Erich Goltze Verlag, pp 403–17

Kleiber M (1961) *Fire of Life*. New York, Wiley, pp 1–453

Lee H Y, Erickson L E and Yang S S (1984a) *J Fermentation Technol* **62**, 341–51

Lee H Y, Erickson L E and Yang S S (1984b) *Biotechnol Bioeng* **26**, 926–35

Lee H Y, Erickson L E and Yang S S (1985) *Biotechnol Bioeng* **27**, 1640–51

Lehninger A L (1982) *Principles of Biochemistry*, Part II. New York, Worth.

Leung J Y C and Wang D I C (1981) *Proceedings 2nd World Congress of Chemical Engineering* **1**, 348–52

Luedeking R and Piret E L (1959) *J Biochem Microbial Technol Eng* **1**, 393–412

Luong J H T and Volesky B (1980) *Can J Chem Eng* **58**, 497–504

Minkevich I G (1980) In: Eroshin V K (ed) *Limitation and Inhibition of Microbial Processes*. Pushchino, USSR, Soviet Academy of Sciences, pp 55–91

Minkevich I G and Eroshin V K (1973) *Folia Microbiol* **18**, 376–85

Oner M D (1985) *Estimation of Yield and Kinetic Parameters in Lactic Acid Fermentations for Yoghurt Cultures*. PhD Dissertation, Kansas State University, Manhattan, Kansas, Ch 2 and 6

Oner M D, Erickson L E and Yang S S (1983) *Biotechnol Bioeng* **25**, 631–46

Oner M D, Erickson L E and Yang S S (1984) *Biotechnol Bioeng* **26**, 1436–44

Patel S A (1981) *A Study of Regularities Associated with Biochemical Processes and Renewable Energy Resources*. MSc Thesis, Kansas State University, Manhattan, Kansas, pp 1–144

Patel S A and Erickson L E (1981) *Biotechnol Bioeng* **23**, 2051–67

Payne W J (1970) *Ann Rev Microbiol* **24**, 17–52

Pirt S J (1965) *Proc R Soc Lond* [*Ser B*] **163**, 224–31

Pirt S J (1975) *Principles of Microbe and Cell Cultivation*. London, Blackwell Scientific, 1–274

Ricci B (1967) *Physiological Basis of Human Performance*. Philadelphia, Lea and Febiger, pp 1–299

Roels J A (1983) *Energetics and Kinetics in Biotechnology*. Amsterdam, Elsevier, Ch 1–5

Samuel W A, Lee Y Y and Anthony W B (1980) *Biotechnol Bioeng* **22**, 757–77

Seagrave R C (1971) *Biomedical Applications of Heat and Mass Transfer*. Ames, Iowa State University, pp 12–13

Solomon B O and Erickson L E (1981) *Process Biochemistry* **16** (No 2), 44–9

Solomon B O, Erickson L E and Yang S S (1983) *Biotechnol Bioeng* **25**, 2683–2705

Solomon B O, Oner M D, Erickson L E and Yang S S (1984) *AIChE Journal* **30**, 747–57

Stouthamer A H (1973) *Antonie van Leeuwenhoek* **39**, 545–65

Stouthamer A H (1977a) In: Haddock B A, Hamilton W A (eds) *Microbial Energetics*. Cambridge, Cambridge University Press, pp 285–315

Stouthamer A H (1977b) *Antonie van Leeuwenhoek* **43**, 351–67

Stouthamer A H (1979) *Int Rev Biochem* **21**, 1–47

Thornton W M (1917) *Philos Mag Sixth Series* **33**, 196–203

Van Dijken T P and Harder W (1975) *Biotechnol Bioeng* **17**, 15–30

Volesky B, Yerushalmi L and Luong J H T (1981) *Proceedings 2nd World Congress of Chemical Engineering* **1**, 268–72

Yang S S, Solomon B O, Oner M D and Erickson L E (1984) *Technometrics* **26**, 355–61

Chapter 3

Calorimetric Techniques

I. Wadsö

3.1 Introduction

Some 200 years ago, the first calorimetric measurements of heat production from animals were performed (Kleiber 1961; Armstrong 1964). About 100 years later, bomb combustion calorimetry was developed and was soon applied to animal and plant materials, food and metabolic products. Results from such measurements and from the continued calorimetric studies of heat production from animals, including humans, were of crucial importance for the development of modern biological energetics. In the 1950s, micro-reaction calorimetry was initiated in biochemistry and biology through the instrument development work by, in particular, Calvet in France and Sturtevant and Benzinger in the USA. However, it was not until the early 1970s that calorimetry started to be used outside a few specialized laboratories in studies of biochemical processes and of cellular systems such as microorganisms and tissue cells, etc.

In recent years instrumental and methodological developments in these areas have been significant and suitable and easy-to-use calorimetric techniques are now available for a wide range of experiments on cellular systems. It is no longer necessary to have extensive experience in calorimetry in order to conduct biocalorimetric work, even if it is performed on a microwatt power level.

3.2 Some Calorimetric Principles and Concepts

For readers who are not acquainted with calorimetry, some principles and concepts of particular importance in the present context will be briefly reviewed. For a general introduction to calorimetry, see Sturtevant (1971). Calorimetric techniques with reference to work in biochemistry and biology have been discussed, e.g. in Brown (1969) and by Wadsö (1970, 1975, 1980), Spink and Wadsö (1976), Martin and Marini (1979) and Spink (1980).

Calorimeters are used for the direct determination of heat quantities, q, and of heat production rates or thermal powers, $P = \mathrm{d}q/\mathrm{d}t$. The thermal power produced in a process is proportional to the rate of the process and calorimeters are therefore also suitable as kinetic instruments, a property which is very often used in work on living systems. Calorimeters are also used for the direct determination of heat capacities, $C = \mathrm{d}q/\mathrm{d}T$, where T is the temperature (K). Microcalorimeters are usually employed in work on cellular systems. This is not a well-defined term, but it normally implies a very sensitive instrument requiring small quantities of material. For instance, a micro-reaction calorimeter can normally be used for precise determination of millijoule (mJ) quantities of heat and the power sensitivity is characteristically on the order of a microwatt (μW). However, it is common, and frequently suitable, to use typical microcalorimeters at a much reduced sensitivity, e.g. in work on small animals or concentrated microbial suspensions, where the thermal power evolved may be several orders of magnitude larger than the sensitivity of the instrument. Ordinary (macro-) solution or reaction calorimeters can also be useful for such work.

Most microcalorimeters are designed as twin, or differential, instruments, where one calorimetric vessel usually contains the reaction system and the other an inactive material. If the differential signal is recorded, contributions from disturbances are expected to cancel. This property is of special importance in cases where the experiments are conducted over long periods of time.

Calorimetric measurement vessels can consist of simple static ampoules or they can have a more complex design, e.g. stirred titration vessels or vessels with two compartments where reagents are mixed by rotation of the calorimeter. Measurements with such vessels are usually performed batchwise and the instruments are sometimes called batch calorimeters, in contrast to (continuous) flow calorimeters. Flow calorimetric vessels typically consist of spiral tubes or ampoules into which reagents are pumped via heat exchangers. Flow-through vessels are often used in experiments with cell suspensions, whereas flow mixing vessels typically are used in chemical thermodynamics. If an object, such as a piece of tissue, is kept in a calorimetric vessel through which liquid or gas can flow, the instrument is often called a perfusion calorimeter.

Several calorimetric measurement principles and many practical designs of calorimeters have found uses in biology and a rather confusing flora of names for the different types of instruments currently exists. From the point of view of measurement principles, calorimeters form two main groups: (essentially) adiabatic calorimeters and heat conduction calorimeters (also called heat leakage, heat flow or heat flux calorimeters).

3.2.1 Adiabatic Calorimeters

In an ideal adiabatic calorimeter there is no net heat exchange between the calorimetric vessel and the surroundings and the heat quantity determined is directly proportional to the observed temperature change, ΔT:

$$q = \epsilon \Delta T \tag{3.1}$$

ϵ is a calibration constant which, in the ideal case, is identical to the heat capacity of the reaction vessel and its content. The thermal power evolved is proportional to the rate of the temperature increase:

$$P = \epsilon \frac{\mathrm{d}T}{\mathrm{d}t} \tag{3.2}$$

In particular, calorimeters used for studies of short chemical reactions or solution processes are often of the 'isoperibol' type (also called 'isothermal jacket calorimeters'). These instruments can be characterized as 'nearly adiabatic'. The reaction vessel is separated from the surroundings by some insulation, e.g. air or a Dewar vessel, but this will not completely prevent heat exchange and in accurate work corrections must be applied (Sturtevant 1971). Even for moderately long reaction periods (of the order of 1 h) the heat exchange term will often cause significant uncertainties. Heat exchange effects can be cancelled out, or reduced, by use of the differential arrangements employed in twin calorimeters.

If the temperature of the calorimetric vessel and its surroundings are kept identical throughout the experiment, no net heat exchange will take place. For endothermic processes this can, in principle, easily be arranged by simultaneous electric compensation in the reaction vessel. For exothermic processes, an 'adiabatic shield' can be inserted between the vessel and the thermostat. The temperature difference between the shield, usually consisting of a metal

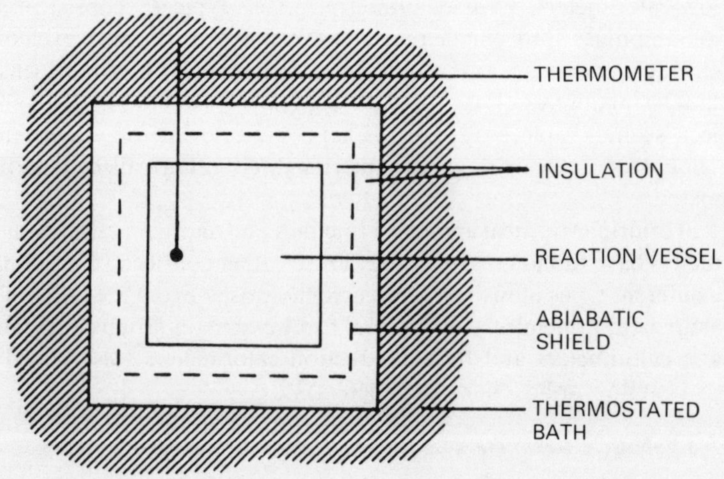

Figure 3.1. Design of an adiabatic shield calorimeter.

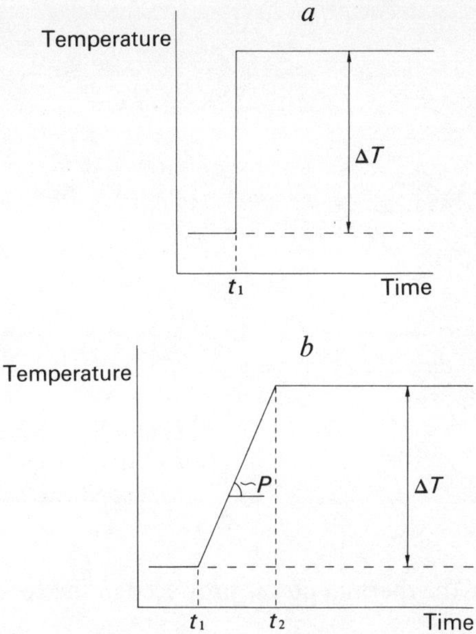

Figure 3.2. Temperature-time curves for an ideal adiabatic calorimeter. *a* A short heat pulse released at time t_1. *b* A constant heat production between t_1 and t_2. The slope of the temperature-time curve is directly proportional to the thermal power, *P*.

envelope, and the calorimetric vessel is kept at zero during the entire measurement by use of an electrical heater. Adiabatic shield calorimeters are important in studies of slow processes and for heat capacity studies.

The principle design of an adiabatic (shield) calorimeter is shown in Figure 3.1. Figure 3.2 shows the temperature-time curve from an adiabatic calorimeter where a constant thermal power is evolved.

3.2.2 Heat Conduction Calorimeters

In the ideal heat conduction calorimeter, heat released is quantitatively transferred from the reaction vessel to a surrounding heat sink, normally consisting of a metal block. With such calorimeters a property proportional to the total heat flow between the calorimetric vessel and the heat sink is measured. Most frequently, the heat flow is recorded by positioning a 'thermopile wall' between the calorimetric vessel and the heat sink (Figure 3.3). The temperature difference over the thermopile will give rise to a voltage signal. In modern thermopile heat conduction calorimeters, the 'thermopile wall' usually consists of semi-conducting thermoelectric plates (Peltier effect plates, which are primarily designed for cooling purposes). The instrument signal is the thermopile voltage, *V*, which in the ideal case is related by the Tian equation

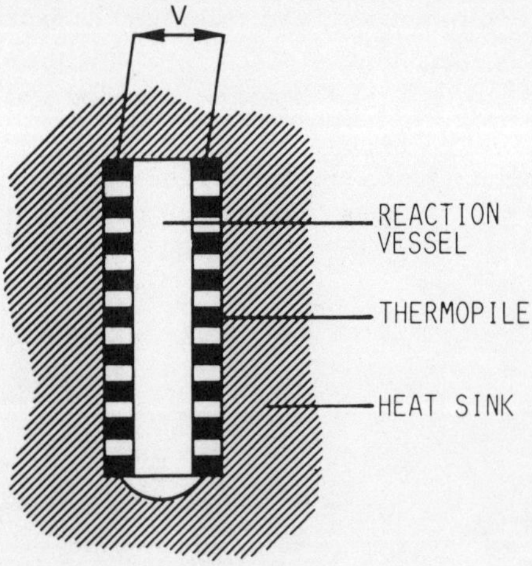

Figure 3.3. Schematic diagram of a section through a thermopile heat conduction calorimeter.

[Equation (3.3)] to the thermal power produced in the vessel:

$$P = \epsilon\left(V + \tau\frac{\mathrm{d}V}{\mathrm{d}t}\right), \tag{3.3}$$

where ϵ is a constant determined by calibration experiment, τ is the time constant for the instrument and $\mathrm{d}V/\mathrm{d}t$ is the time derivative of the voltage signal. The value for τ is proportional to the heat capacity of the reaction vessel and inversely proportional to the thermal conductivity of the thermopile. In cases where there are significant temperature gradients in the reaction vessel, two or more time constants should be used (Randzio and Suurkuusk 1980).

If a constant power is produced, V will reach a steady-state value, V_s. The power released in the vessel is then quantitatively balanced by the heat leakage to the surroundings and Equation (3.3) will be reduced to:

$$P = \epsilon V_s \tag{3.4}$$

i.e. the power is directly proportional to the voltage signal from the thermopile. To a good approximation, the simple Equation (3.4) can be applied to processes where the power changes slowly compared to the time constant, which is often the case in processes involving living cells. In other cases, thermopile heat conduction calorimeters will give voltage-time curves which are significantly distorted compared to the true kinetic curve. To obtain precise kinetic information under such conditions, it is necessary to reconstruct the power-time curve from the thermopile output data (Randzio and Suurkuusk 1980). This can now be performed automatically by use of a computer connected on-line to the calorimeter, but as yet it is rare to find that signals from thermopile conduction calorimeters are corrected to the true kinetic curves. For comparatively rapid rate changes such as those sometimes observed in microbial growth experiments, it is likely that the corrected and uncorrected curves will be significantly different.

In work with a microtitration calorimeter, Gill and coworkers have recently demonstrated (McKinnon *et al* 1984) that the time constant can be reduced significantly through a computer-controlled electrical compensation technique.

Figure 3.4 shows typical voltage-time curves from this type of calorimeter. Integration of Equation 3.3 or 3.4 between t_1 and t_2 (Figure 3.4b) will give:

$$q = \epsilon \int_{t_1}^{t_2} V \, dt \tag{3.5}$$

or

$$q = \epsilon A \tag{3.6}$$

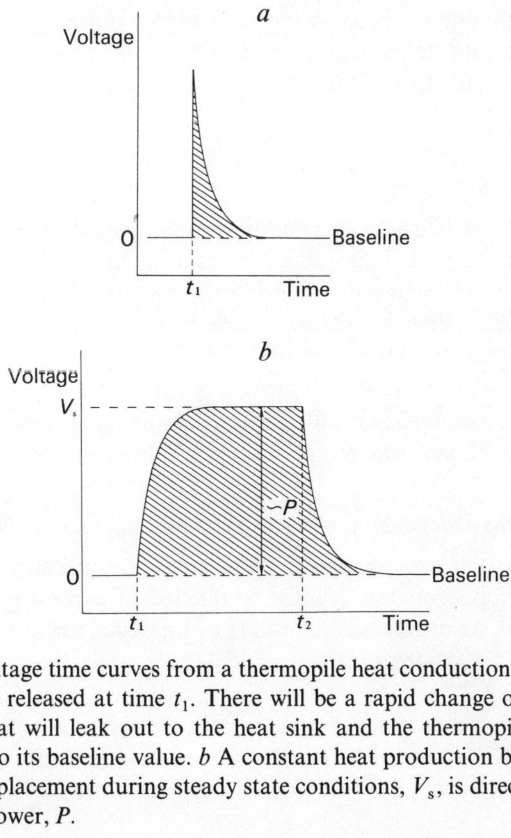

Figure 3.4. Voltage time curves from a thermopile heat conduction calorimeter. *a* A short heat pulse released at time t_1. There will be a rapid change of the thermopile voltage. The heat will leak out to the heat sink and the thermopile signal returns asymptotically to its baseline value. *b* A constant heat production between t_1 and t_2. The baseline displacement during steady state conditions, V_s, is directly proportional to the thermal power, P.

where A is the surface area under the voltage-time curve. If the integration is extended to the time where the voltage signal has returned to the baseline, it is thus not necessary to know the time constant in order to determine q.

The sensitivity of a heat conduction calorimeter can be expressed as the quotient between the values for the thermopile voltage and the power (unit: $V\,W^{-1}$), or as the ratio between the voltage-time integral and the heat quantity (unit $V\,s\,J^{-1} = V\,W^{-1}$). The sensitivity value and the calibration constant, ϵ, are thus independent of the heat capacity of the system. Therefore, in an experiment with a given cell suspension, the value for the calorimetric signal ideally will increase in direct proportion to the amount (volume) of cell suspension. This is in contrast to the case of adiabatic calorimeters, where it is the change in temperature which is measured. In the ideal case, where the heat capacity for the vessel material can be neglected, the measured temperature change or the rate of temperature change will not increase if the amount of a given heat-producing suspension is increased.

By the use of a thermopile, it is also possible to employ the Peltier effect principle for an active transport of heat released in the process. The cooling power produced $-P$, is proportional to the electrical current, I, through the thermopile. In practice, this cooling power will be superimposed on the joule heating power produced throughout the circuit and will be partly transmitted to the thermopile junctions (corresponding to an effective resistance, r). The cooling power is thus given by:

$$-P = k(I - rI^2) \tag{3.7}$$

where k is a constant. Calorimeters of this kind are often called thermoelectric heat pump calorimeters. Normally the reaction heat is balanced with the cooling power (which is measured) and the temperature of the calorimeter will thus be constant. These calorimeters are therefore also sometimes called isothermal calorimeters. However, in this context the term can be confusing, as thermopile heat conduction calorimeters are frequently used under very nearly isothermic conditions and the term 'isothermal calorimetry' is frequently used as a contrast to 'temperature scanning calorimetery'.

3.2.3 Isothermal Calorimetry and Temperature Scanning Calorimetry

For obvious reasons, measurements of different physical and chemical properties of cellular systems are usually performed at a constant temperature. Thus isothermal, or nearly isothermal, calorimetry is the most important in work on cells. Adiabatic calorimeters cannot be used to measure truly isothermal processes, but ΔT can be kept small, often of the order of 0·1 °C or less, if sensitive calorimeters are used. For a thermopile heat conduction calorimeter using semi-conducting thermopiles, the ΔT over the thermopiles is very small, typically of the order of 10^{-4} °C, when the power is 100 μW.

Temperature-scanning calorimeters are usually constructed as differential instruments (DSC = differential scanning calorimeter), where the experimental temperature changes continuously during the measurement. These instruments, which are heat capacity calorimeters, are of great importance in biothermodynamics, particularly for the characterization of thermally induced transitions of biopolymers and aggregates such as membrane materials (Privalov 1980; Bach and Chapman 1980; Edsall and Gutfreund 1983). However, DSC is of limited interest in studies of metabolizing cellular systems and will not be treated in this chapter. It is sometimes of interest to study the thermal power for a cellular system at different temperatures from which the (apparent) activation energy can be calculated (Monti and Wadsö 1976). However, such measurements are best performed under isothermal conditions.

3.3 Some Calorimetric Designs

Calorimetric measurements on cellular systems are normally conducted within a narrow and convenient range of temperatures and at atmospheric pressure. Realistic requirements for precision and accuracy are significantly lower than those usually required in chemical thermodynamic studies. On the other hand, special precautions must be taken to ensure that the physiological conditions are suitable and well-characterized: p_{O_2}, pH, sedimentation and adhesion effects, changes in viability, etc. Such parameters are often difficult to handle without introduction of thermal disturbances, which might lead to random or systematic errors, particularly in cases where the heat production rate is low and the measurement is extended over a long period of time.

It is characteristic for the calorimetric field that many non-commercial instruments are in use. Some of these have been designed to meet special experimental problems, but in many cases it appears that homebuilt instruments have been used for economic reasons. It can be a very time-consuming task to build a calorimeter and to develop adequate working procedures. It might therefore be unwise to start work in biocalorimetry by designing a sophisticated instrument rather than buying commercially available equipment, even if the equipment has to be supplemented or somewhat modified in order to suit some special needs. However, as will be illustrated below, there are certainly problems in biological calorimetry which can be studied successfully by use of simple instruments which can be assembled in most laboratories.

The following discussion is meant to cover recent calorimetric designs suitable for work on metabolizing cellular systems, including various tissues, organs, and small animals. Some well-established instruments, much used in these fields, will also be included.

3.3.1 Adiabatic and Isoperibolic Calorimeters

Truly adiabatic calorimeters, which have played a very important role in chemical thermodynamic studies of slow processes and in temperature scanning calorimetry, are currently not much used in work on living cells.

Kresheck (1974) used an adiabatic reaction and solution calorimeter in work on mouse leukaemia cells. In that instrument, the temperature of the thermostatic bath was monitored to follow that of the reaction vessel and it thus acted as the equivalent of an adiabatic shield (Benjamin 1963).

Figure 3.5 shows a simple Dewar vessel calorimeter working after the same principles. It was used for studies of the heat produced in nitrogen fixation by soybean nodules (Heytler and Hardy 1984). A thermocouple pair was used to monitor the increase of the bath temperature. The temperature of the vessel was also measured by means of a thermocouple. During the measurements, which were extended over a period of about 1 h, large quantities of heat were evolved and the temperature increased by as much as 2–5 °C; the non-isothermal conditions called for a correction of the measured reaction rate. (A value of $Q_{10} = 2$ was assumed for all metabolic processes.)

Loesberg *et al* (1982) have reported work on rat hepatoma cells performed with a very simple isoperibol Dewar vessel calorimeter (Figure 3.6). The rather

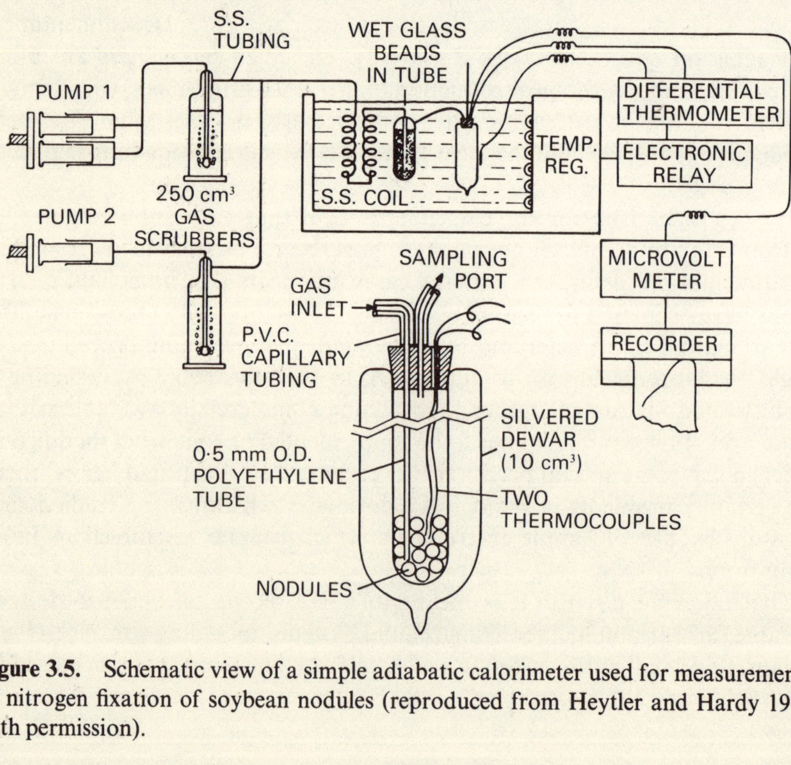

Figure 3.5. Schematic view of a simple adiabatic calorimeter used for measurements of nitrogen fixation of soybean nodules (reproduced from Heytler and Hardy 1984 with permission).

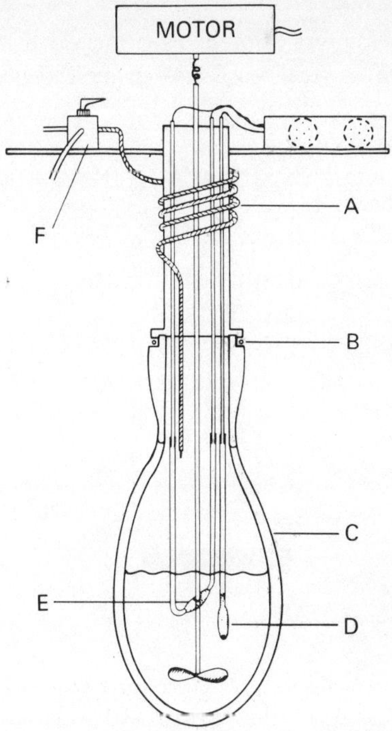

Figure 3.6. A very simple isoperibolic calorimeter used for measurements of suspended hepatoma cells. The vessel is immersed in a precisely thermostated water bath: A stainless steel tubing (inlet second component); B Viton O-ring; C Dewar vessel; D thermistor; E calibration heater; F Hamilton 3-way valve (reproduced from Loesberg *et al* 1982 with permission).

large vessel, volume 100 cm^3, is fitted with a sample inlet tube and a stirrer. The temperature is measured with a thermistor. For a typical experiment, 2×10^8 cells were used and the measurement accuracy was estimated to be 2–5%. The power level was then about 3 mW. The quantity of cells is much larger than would normally be used with a typical microcalorimeter, but the authors have demonstrated that very simple equipment can be used for measurements on tissue cells. For this type of calorimeter, in contrast to heat conduction calorimeters, the calorimetric signal is proportional to the quotient between the total power evolved and the heat capacity of the vessel and its contents. Ideally, one would thus not get large decreases in the sensitivity if a smaller volume of the same cell suspension was measured (in a smaller vessel). However, difficulties with practical factors such as irregularities in the heat exchange effects and friction effects would be expected to increase.

Tronac, Inc. (Orem, Utah) has on its program small volume titration calorimeters of the isoperibol type which have found use in work on cell suspensions (Hansen *et al* 1980; Gonda *et al* 1985).

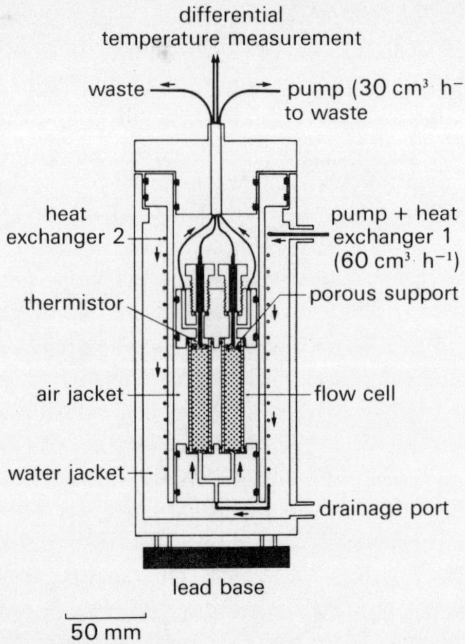

Figure 3.7. A simple twin flow calorimeter for studies of cells attached to a solid support contained in the flow cells (reproduced from Lock and Ford 1983 with permission).

Another comparatively simple isoperibol calorimeter (Figure 3.7) was built by Lock and Ford (1983). The instrument is a twin flow (or perfusion) calorimeter used in measurements of aquatic microorganisms attached to a solid support, e.g. sand or glass beads kept in flow cells. The instrument, constructed mainly from Perspex (Plexiglass), used thermistors as temperature sensors. The detection limit was estimated to be 3 μW. This instrument is one of the same type as those developed by Mosbach and coworkers (Danielsson *et al* 1981) and called 'enzyme thermistors' (an unfortunate term!). They have been used successfully in analytical work with immobilized enzymes and also with immobilized microorganisms (Mattiasson *et al* 1977).

The flow calorimeter designed by Picker *et al* (1971) is a differential instrument which uses a modulated thermal detector device requiring a single thermistor as sensor. This instrument was designed for use as a mixing calorimeter, but has been used, in a slightly modified form, in work on microbial systems in connection with a study on waste water treatment (Fortier *et al* 1980). The Picker calorimeter is commercially available from Sodev, Inc. (Sherbrooke, Quebec, Canada) and Setaram (Caluire, France).

3.3.2 *Heat Conduction Calorimeters*

For work in the fields treated in the present volume, thermopile heat conduction instruments manufactured by Setaram (Caluire, France) and LKB Produkter (Bromma, Sweden) seem to have been the most widely used among commercial instruments. One of the best known types of heat conduction calorimeters is the Calvet microcalorimeter, developed by E. Calvct in Marseille and based on the early design by Tian (Calvet and Prat 1956, 1963). These calorimeters have played an important role in many areas of application, not the least in biology, and the pioneering work performed by Prat (Calvet and Prat 1963) on living systems has inspired many workers to initiate further exploration of this field. Figure 3.8 shows schematically a standard Calvet-type calorimeter marketed by Setaram. This instrument is suitable for measurements at the microwatt level and has a good long-term stability. A large aluminium block serves as the heat sink for two (or four) calorimetric units (one is indicated in Figure 3.8a), which are normally used as a differential system(s). In each unit, a thin-walled metal cylinder defines the space for the calorimetric vessel and is surrounded by a wire-wound thermopile (Figure 3.8b) which is in contact with the heat sink. The reaction vessels often consist of simple tubes with stoppers and reaction volume 15–100 cm^3, but many different vessels for special purposes (e.g. injection, stirring, introduction of light) have been described (Calvet and Prat 1956, 1963; Spink and Wadsö

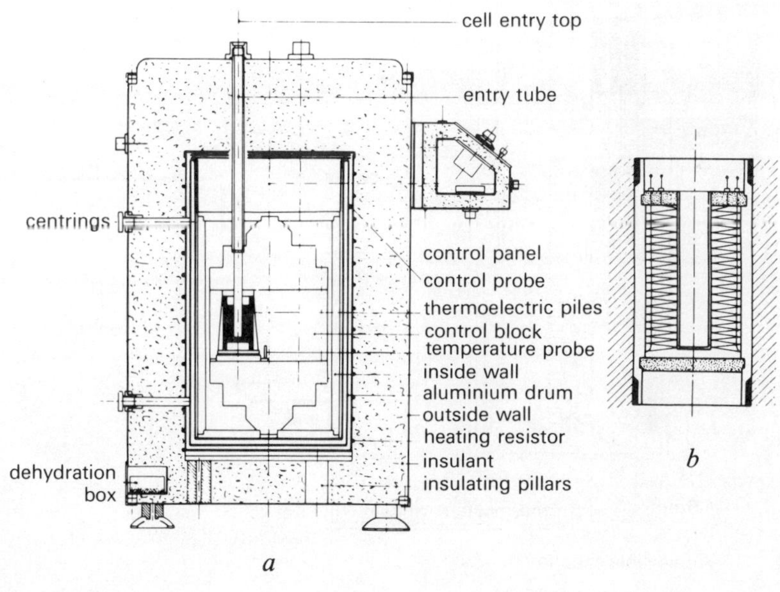

Figure 3.8. *a* A schematic diagram of a twin microcalorimeter of the Calvet type. One of the calorimetric units is indicated. *b* Detector arrangement (courtesy of Setaram, Caluire, France).

1976). Some specialized vessels are commercially available. Calvet-type instruments where the whole block can be rotated in order to achieve mixing have also been reported (Evans 1969). For some applications, a separate thermopile is used for Peltier effect cooling (Calvet and Prat 1956, 1963).

More recently, Setaram has introduced a new multipurpose calorimetric system, 'Bio-DSC', which is also based on the heat conduction principle. This instrument is designed for temperature scanning experiments as well as for isothermal batch and flow calorimetry (Benoist and Pithon 1984). The arrangement of the calorimeter is shown schematically in Figure 3.9. The temperature is controlled by a flow of thermostated liquid. The vessel volume is 1 cm^3. No detailed report on the design and the performance of the instrument seems to have been made.

The 'Bioflux' microcalorimeter (Thermanalyse, Grenoble, France) is a rather simple twin heat conduction calorimeter of the Calvet type. It is used with insertion vessels with a maximal volume of 25 cm^3. Short-time baseline stability is stated to be $\pm 2\,\mu W$. This instrument has been used in studies of skin cells attached to a plastic foil (Schaarschmidt and Reichert 1981).

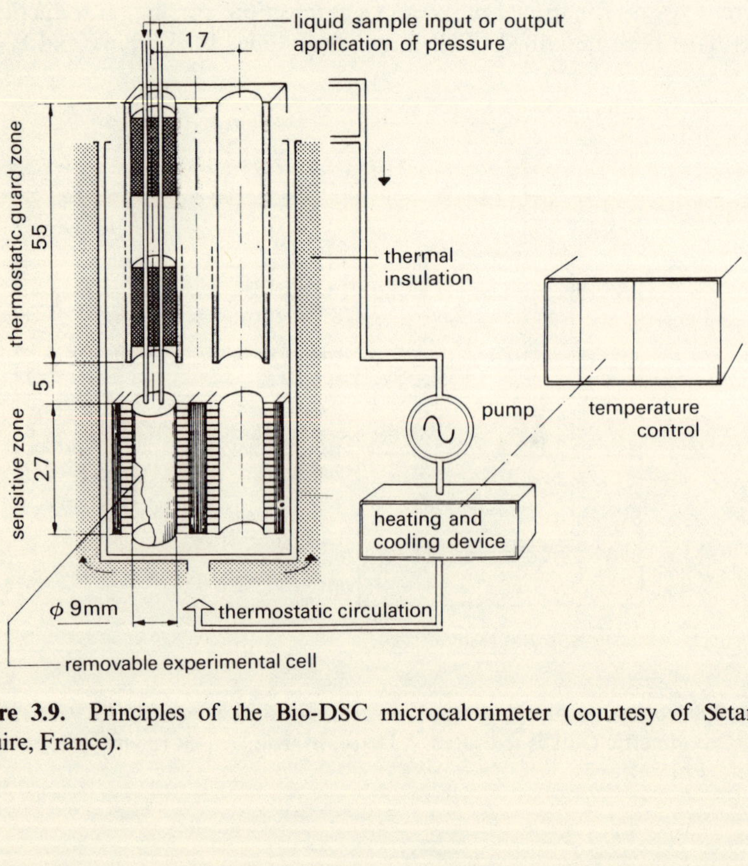

Figure 3.9. Principles of the Bio-DSC microcalorimeter (courtesy of Setaram, Caluire, France).

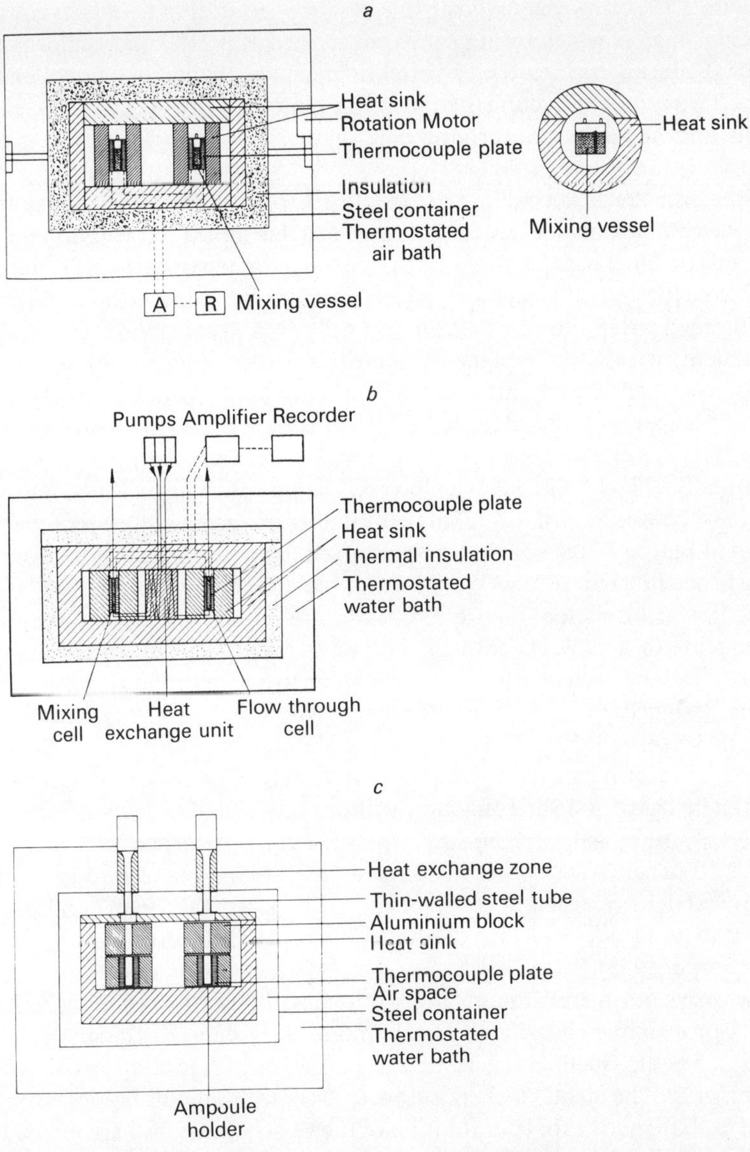

Figure 3.10. Schematic diagrams of some twin heat conduction microcalorimeters using semiconducting thermocouple plates (prototypes to the LKB 2107 microcalorimetric system). *a* Rotating batch microcalorimeter (after Wadsö 1968); *b* Flow microcalorimeter (after Monk and Wadsö 1968); *c* Ampoule microcalorimeter (after Wadsö 1974).

Figure 3.10 shows schematically the design of a series of twin heat conduction calorimeters which form prototypes to the LKB 2107 microcalorimetric system. Calorimetric vessels, or vessel holders, are shaped like boxes or thin plates. Two semi-conducting thermocouple plates are interspaced between the vessels and metal blocks forming part of the heat sink. The useful power sensitivities are about 1 μW.

In the batch reaction calorimeter (Figure 3.10a) (Wadsö 1968), reagents in the two-compartmented reaction vessel can be mixed by rotation of the calorimetric block. Jarret *et al* (1979), working on hepatocytes, introduced a modified experimental technique which significantly reduced the equilibrium time needed after introduction of the cells into the reaction vessel. This instrument, which was designed primarily for biochemical binding experiments, has recently been fitted with a programmable titration unit, by means of which reagents can be added to the vessels during a measurement (Beezer *et al* 1982).

Prosen (1973) developed a single-vessel batch calorimeter which has been used in some studies with cells. This calorimeter has a mixing vessel of the type shown in Figure 3.10a, which can be removed from the instrument during the filling procedure (Spink and Wadsö 1976; Martin and Marini 1979).

The flow calorimeter (Figure 3.10b) is quite similar in design to the batch instrument. In a flow experiment, one or two flows of liquid are pumped through the heat exchange unit and from there to one of the flow vessels. These can be designed as mixing vessels where two flows are mixed, or as flow-through vessels (Figure 3.11). It is often suitable to equip a twin flow (or perfusion) calorimeter with two different vessels (Figure 3.10b), which can give the instrument a dual function without any significant loss of the twin properties. Flow calorimeters used in work on cell suspensions are often connected in series with other instruments such as oxygen electrodes, meters for optical density, fraction collectors or valve systems for sequential sample collection or introduction (Eriksson and Wadsö 1971; Brettel *et al* 1980, 1981; Hoffner *et al* 1985).

The twin calorimeter (Figure 3.10c) is used with insertion vessels which can consist of a simple closed cylindrical ampoules (volume 1–10 cm^3) or more complex vessels, such as the vessel for perfusion of liquid or gas shown in Figure 3.11g. The metal block, positioned above the ampoule holder, serves as a heat exchange zone for equilibration of the vessels when they are introduced into the measurement position (the block is in thermal contact with the main heat sink). It is also used for equilibration of the flow in perfusion vessels. This type of calorimeter has been used extensively in work with blood cells and other animal cells (p. 131), and with bacterial systems including soil (Wadsö 1980).

The LKB 'sorption' microcalorimeter, based on the design by Kusano *et al* (1971), is very similar to the instruments shown in Figure 3.10. Gnaiger (1983a, b) used this calorimeter in work on small aquatic animals. It functioned as a perfusion instrument (Figure 3.12), connected in series with a twin flow respirometer (a polarographic oxygen sensor manufactured by Cyclobios, Innsbruck, Austria). The adsorption column of the calorimeter was used as the animal chamber through which the medium was perfused before reaching the respirometer. For work on similar animal systems, Knudsen *et al* (1983) used an LKB ampoule calorimeter (Figure 3.10c), together with a perfusion vessel of the type shown in Figure 3.11g. The flow from the vessel was led to a 0·1 cm^3 (100 μl) chamber holding an oxygen electrode. Kemp (1975, 1980) used a modified version of the LKB sorption cell in flow calorimetric work on tissue cells in monolayers, see Figure 3.11f. The cells were attached to glass beads placed in the adsorption vessel. Gordon and Millero (1983) used the sorption calorimeter for work on bacteria attached to polyacrylamide beads placed in the adsorption column.

Figure 3.13 shows schematically a 4-channel microcalorimeter (Suurkuusk and Wadsö 1982) which forms the prototype of the LKB 'BioActivity Monitor'. A thermostated water bath can hold up to four independently operated twin calorimeters ('channels'). The temperature stability of the bath is $\pm 1 \times 10^{-4}$ °C over long periods of time (days), if the room temperature does not fluctuate more than ± 1 °C. Figure 3.13b shows the design of a twin calorimetric channel used with cylindrical insertion vessels, diameter 14 mm.

Above the ampoule holders the bores are narrowed in order to give a close fit with the calorimetric insertion vessels and thus serve as heat exchange zones. In another type of channel, the twin calorimetric unit is fitted with two permanently installed flow vessels used for mixing and flow-through experiments, respectively. Both vessels consist of gold tubes wound on threaded aluminium tubes. Except for the thread, the aluminium tubes are closely similar to the ampoule holders shown in Figure 3.13(h) and can thus also serve as holders for insertion vessels. Several types of simple insertion vessels are used: 1, 3 or 5 cm^3 cylindrical steel ampoules or disposable 3 cm^3 glass vials. The long-time baseline drift and the maximum short-term deviation from the baseline for such vessels are typically less than 0·1 μW (Figure 3.14). A 20 cm^3 ampoule used with a single calorimetric channel is significantly less stable.

Recently a more complex insertion vessel (Figure 3.15) has been developed (Görman Nordmark *et al* 1984) for cells in suspension or attached to a solid support (polystyrene film, microcarriers), for small pieces of tissues and small animals (volume < 1 cm^3). Experiments can be performed with gas phase present or with the vessel completely filled with medium. The contents can be stirred and liquid can be perfused through the vessel. Small quantities of reagents can be added to the sample compartment through injection needle(s) during the measurement and stepwise titrations can be performed by an automatic procedure. The baseline stability is typically $\pm 0{\cdot}3$ μW over 10 h (stirring rate 50 rpm, perfusion rate $\leqslant 20$ cm^3 h^{-1}). Gas can also be pumped through the system, with some increase of the noise level. Different specialized

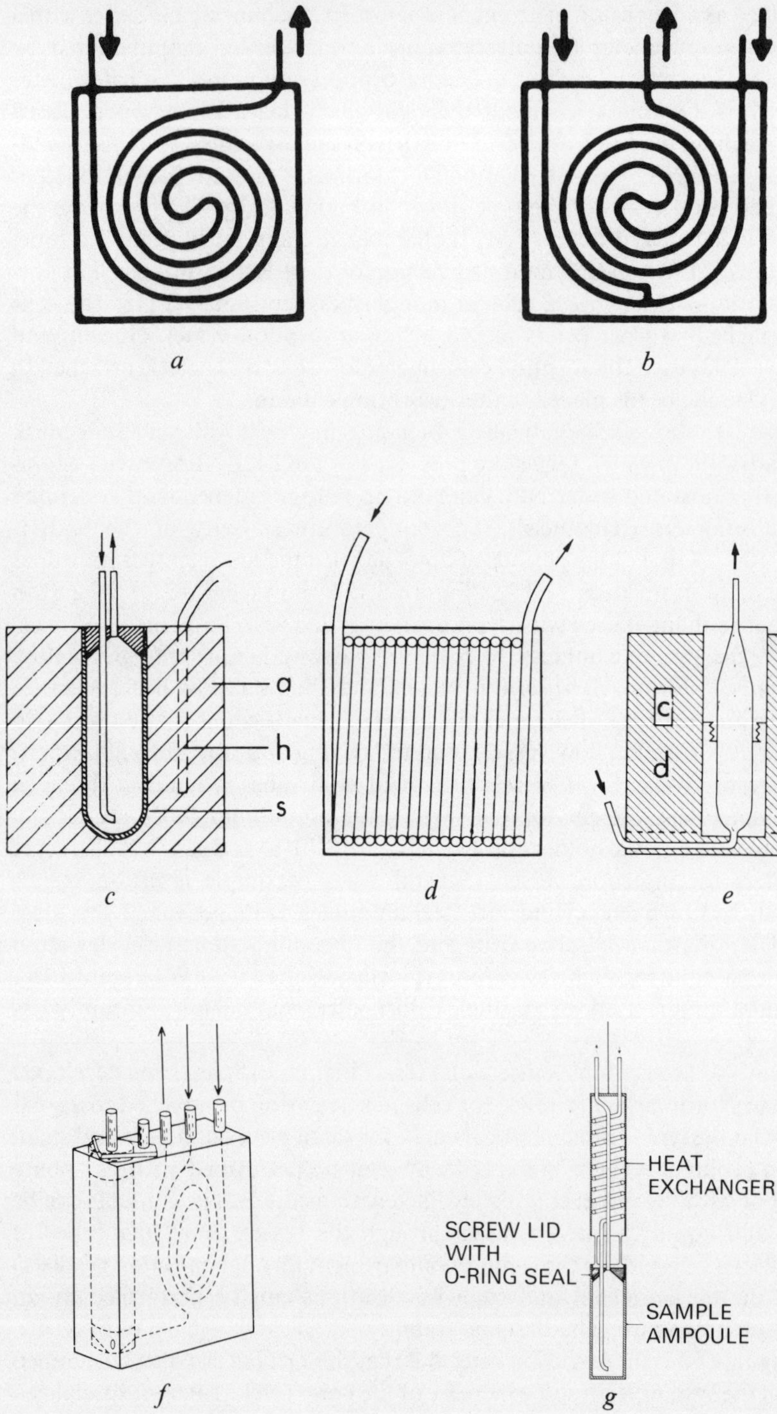
a
b
a
h
s
c
d
c
d
e
HEAT
EXCHANGER
SCREW LID
WITH
O-RING SEAL
SAMPLE
AMPOULE
f
g

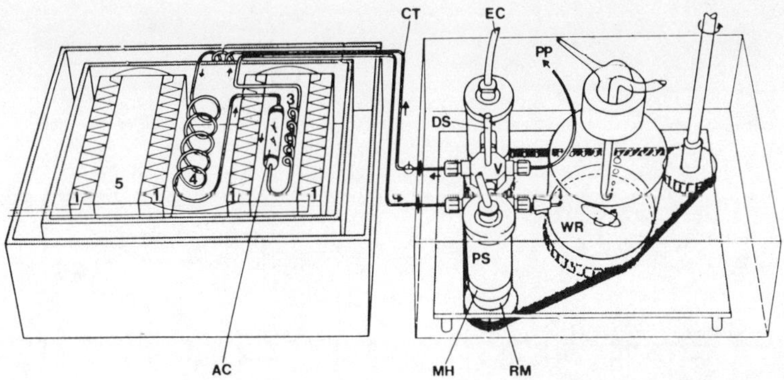

Figure 3.12. A twin-flow respirometer (Cyclobios) connected to the LKB (flow) sorption calorimeter and used in experiments with small aquatic animals. AC, the sorption vessel used as an animal chamber; CT, capillary tube; DS, drive shaft for switching the two four-way valves; EC, electrode cables of the polarographic oxygen sensor; MH, magnet housing; PP, outlet Teflon tubing connected to the peristaltic pump; PS, polarographic oxygen sensor sleeve; RM, rotating magnet; V, four-way microvalve; WR, water reservoir for gas equilibration; 1, thermocouple plates; 3,4, heat exchange coils; 5, static reference plate (reproduced from Gnaiger 1983b with permission).

stirrers and sample holders have been tested (Figure 3.16). The turbine stirrer (Figure 3.16b) can be used with microcarriers and other relatively heavy suspensions. (It also works well without the propeller blades indicated in the Figure.) The cage sample holder (Figure 3.16c) has recently been used for measurements of human muscle fibres (Fagher *et al* 1985). During such measurements, the cage is rotated while a flow of medium is perfused through the vessel.

Figure 3.11. Some flow calorimetric vessels used with instruments of the types shown in Figure 3.10 *b* and *c*. *a,b* Flow vessels consisting of spiralized gold tubes soldered between copper plates (LKB, Bromma, Sweden). *a* is a flow-through vessel and *b* is a mixing vessel. *c,d,e,f* Flow-through vessels used for liquid flow or mixed (segmented) gas-liquid flow. *c* is a cylindrical stainless steel vessel (s) inserted into a hole in an aluminium plate (a). (h) indicates a calibration heater (Eriksson and Wadsö 1971). *d* consists of a teflon tube which is moulded into tin in a thin-walled brass box (Wadsö 1980, cf. Kemp 1980). *e* is a cylindrical steel vessel inserted into an aluminium plate (d), which also houses a calibration heater (c) (Gustafsson and Lindman 1977). *f* is a cylindrical glass vessel (LKB sorption vessel, cf. the animal chamber in Figure 3.12) inserted into a metal plate which also contains a heat exchange coil connected to the glass vessel. Suitable choice of inlet and outlet allows upward or downward flow in the vessel (Kemp 1975). *g* Perfusion (gas or liquid) vessel used with the calorimeter shown in Figure 3.10 *c* (Wadsö 1974, cf. Knudsen 1983).

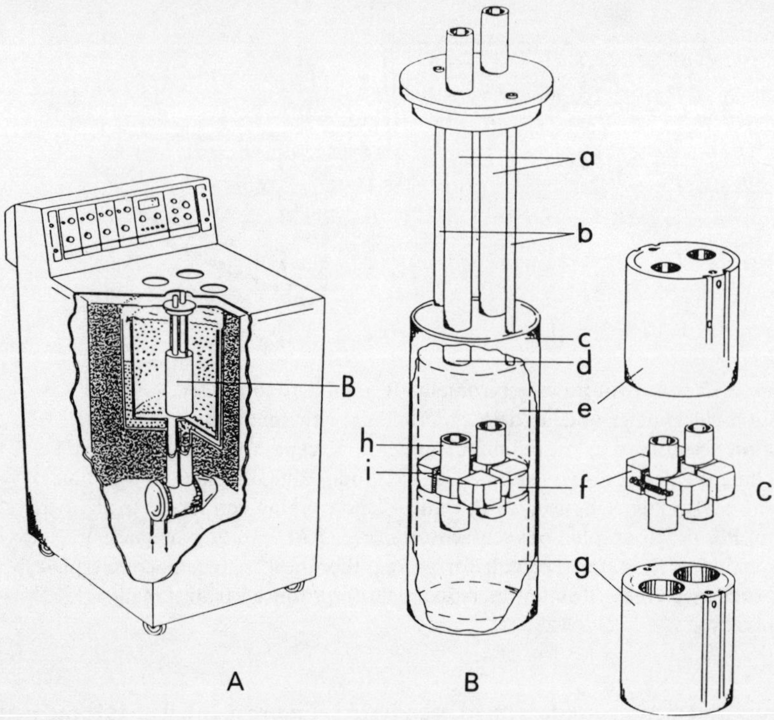

Figure 3.13. 4-channel microcalorimeter (prototype for LKB's 'BioActivity Monitor'). A Cutaway view of the instrument shown with one channel (B) positioned in the water bath. C is the twin calorimeter unit. a,b thin-walled steel tubes; c, steel cylinder; d, plastic (Ryton) tube; e,g, aluminium bolts serving as main heat sinks; f, aluminium block; h, aluminium tube (holder for insertion vessel); i, thermocouple plate (after Suurkuusk and Wadsö 1982).

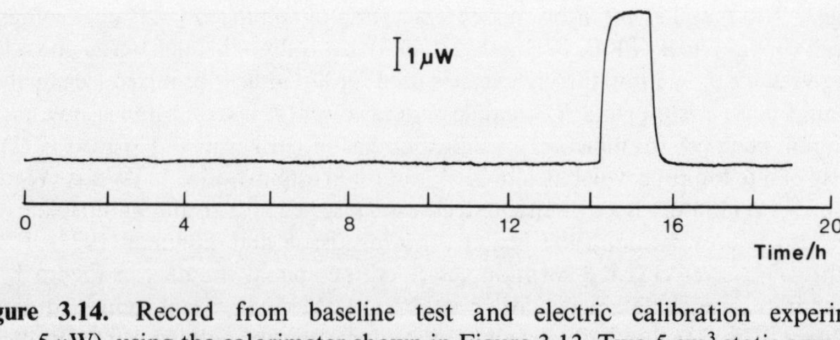

Figure 3.14. Record from baseline test and electric calibration experiment ($P = 5\,\mu$W), using the calorimeter shown in Figure 3.13. Two 5 cm^3 static ampoules were charged with water (reproduced from Suurkuusk and Wadsö 1982 with permission).

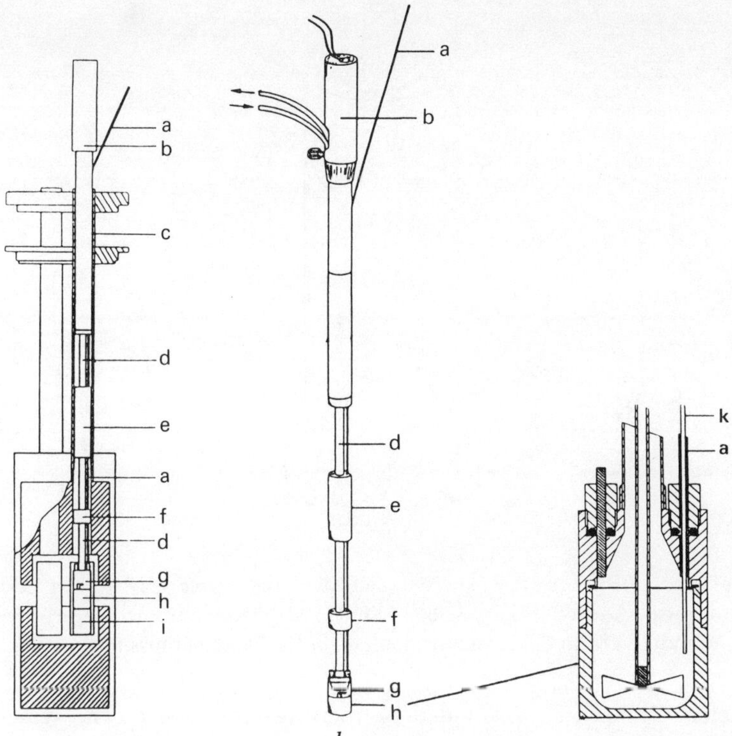

Figure 3.15. Insertion vessel used for perfusion and titration experiments with the instrument shown in Figure 3.13. *a* Simplified diagram of the vessel in measurement position; *b* the vessel; *c* section through the sample compartment (1 cm^3 sample cup). a, injection tube; b, stirrer motor; c, plastic tube; d, steel tube; e, f, brass bolts; g, upper part of sample compartment; h, sample cup; i, aluminium support; k, injection needle (after Görman Nordmark *et al* 1984).

Lovrien and coworkers have reported the design of several typical thermopile heat conduction calorimeters for work with microorganisms, different tissues and insects. Figure 3.17 shows a twin perfusion calorimeter (Anderson and Lovrien 1979) used for work with plant tissues. Thermally equilibrated medium flows through thermostating coils, which are in contact with the heat sink, before reaching the cylindrical sample chambers (diameter 15 mm, length 50 mm). Flow rate is typically 60 $cm^3 h^{-1}$. The unit (Figure 3.17) is surrounded by aluminium shells which are thermostated by use of Peltier effect plates.

Figure 3.18 shows schematically a twin instrument for studies of insects, in particular in assays of their response to airborne compounds such as pheromones (Kurtti *et al* 1979; Lovrien *et al* 1981). The two flat glass vessels,

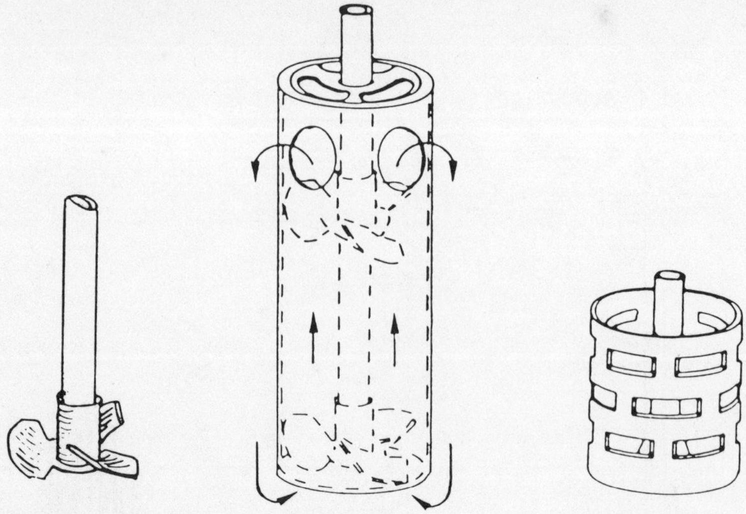

Figure 3.16. Stirrers which can be attached to the stirrer shaft of the perfusion titration vessel shown in Figure 3.15. *a* Simple propeller; *b* 'turbine' stirrer; *c* rotating cage (reproduced from Görman Nordmark *et al* 1984 with permission).

diameter 35 mm, are surrounded by thermocouple plates. The vessels are connected by a tube which can be closed by a stopcock. Thermostated air inflow and outflow (ca. 1–2 $cm^3 min^{-1}$) are provided for each vessel. The inner section of the calorimeter shown in Figure 3.18 is surrounded by insulation and aluminium shells thermostated by use of Peltier effect plates.

Lovrien *et al* (1980) have also described rotating batch calorimeters where the liquids in compartmented vessels are mixed by rotation of the calorimetric block. Figure 3.19 shows a twin calorimeter with glass mixing vessels (volume ca. 2 cm^3) used in aerobic or anaerobic microbial experiments (Lovrien *et al* 1980). Later versions of this instrument hold several vessels which are mounted in a row on a keel block (Hammerstedt and Lovrien 1983). A five-vessel instrument of this type, manufactured by Microcalorimeters, Inc., St. Paul, Minnesota, was used by Innskeep and Hammerstedt (1983) in work on bovine sperm.

Ito *et al* (1984) used a two-channel heat conduction calorimeter in work on ascites carcinoma cells. Cell suspensions and reagents can be injected into two stirred reaction vessels, each in contact with a thermocouple plate. The voltage from the two calorimetric units is compared with that from a reference thermopile. The instrument, which has a rather low sensitivity, was manufactured by Rhesca, Tokyo, Japan.

Fujita and coworkers have reported the design of a rotating batch calorimeter which was equipped with vessels specially designed for studies of aerobic

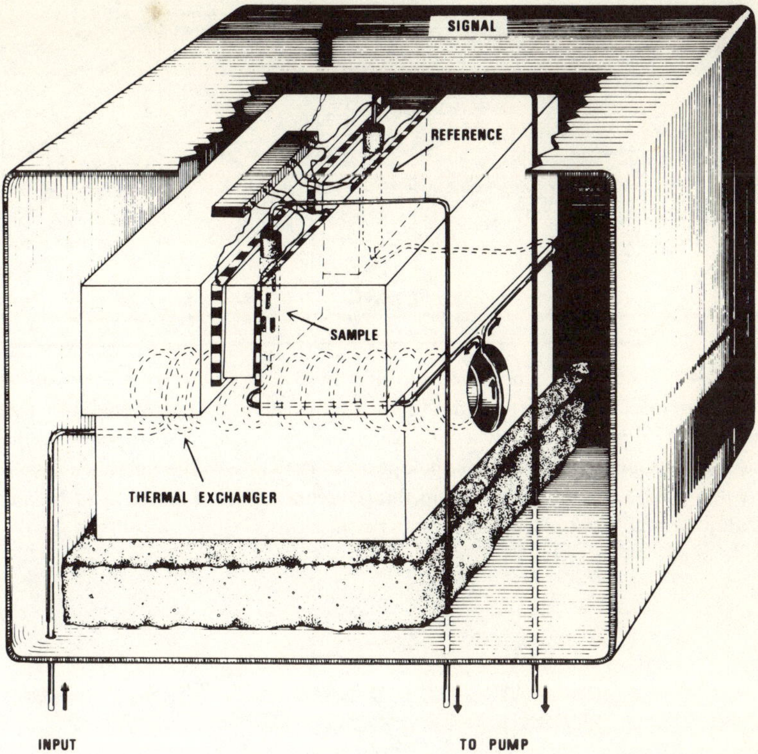

Figure 3.17. The interior of a twin perfusion calorimeter used with, e.g. plant tissues. Sample vessels are inserted into aluminium plates which are surrounded by thermocouple plates. Incoming solutions are thermostated in a heat exchange coil which is in thermal contact with the heat sink (reproduced from Anderson and Lovrien 1979 with permission).

microbial growth (Fujita *et al* 1976; Nunomura and Fujita 1982). A more recent version of this instrument was described by Kobayashi *et al* (1985). It can be used in either batch or flow mode (Figure 3.20). This type of batch calorimeter is produced by Tokyo Riko Co., Tokyo.

It can be argued that there is no reason to use high sensitivity microcalorimeters with small volume vessels in work on growth processes for microbial systems, where the material is inexpensive and the thermal power evolved is high. Such arguments led Takahashi (1973) to build a twin thermopile heat conduction calorimeter where the volume of the stirred vessels was 300 cm^3. [Also see the work of Mou and Cooney (1976) who used a regular 7 dm^3 fermentor as a calorimeter.]

The instrument reported by Ishikawa *et al* (1981) was specially designed to handle the problem of efficient oxygen supply to microbial cultures. The

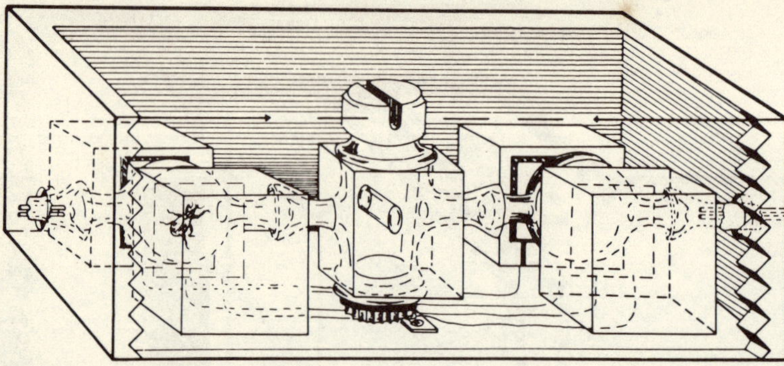

Figure 3.18. The measurement section of a twin calorimeter used for assay of the effects on an insect of airborne compounds. Flat glass vessels, surrounded by thermocouple plates, are connected by a tube which can be closed by a stopcock. When the passageway is open, the insect has the option of crawling to the reference vessel. The active compound can be deposited in the reference vessel or brought in by air flow (reproduced from Lovrien *et al* 1981 with permission).

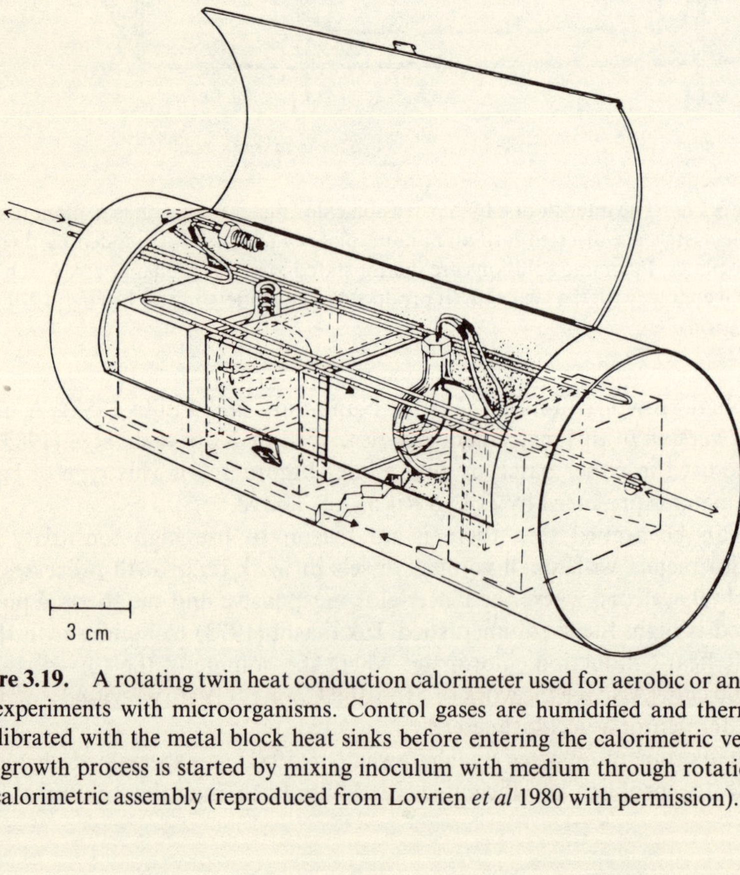

Figure 3.19. A rotating twin heat conduction calorimeter used for aerobic or anaerobic experiments with microorganisms. Control gases are humidified and thermally equilibrated with the metal block heat sinks before entering the calorimetric vessels. The growth process is started by mixing inoculum with medium through rotation of the calorimetric assembly (reproduced from Lovrien *et al* 1980 with permission).

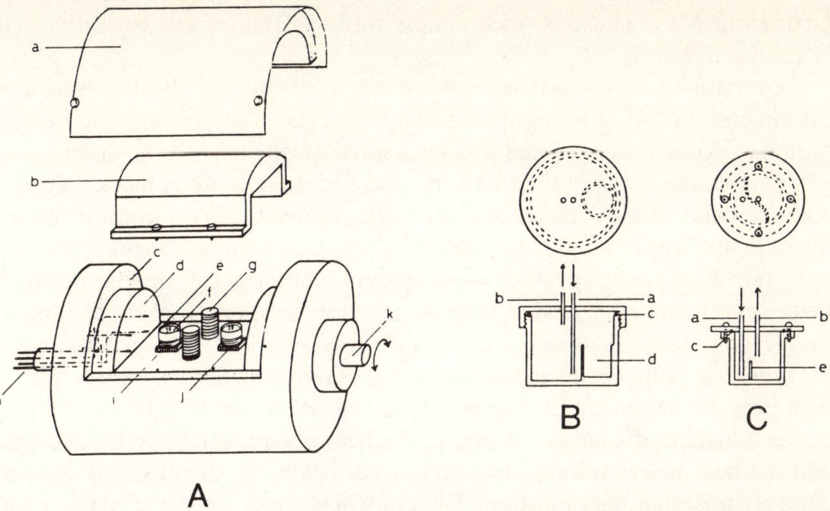

Figure 3.20. A rotating twin heat conduction calorimeter designed for use with microbial systems with high oxygen consumption and large or heavy cells. A The instrument can be used with batch vessels or with flow vessels. a, styrofoam insulation lid; b, aluminium lid; c, styrofoam; d, heat sink; e, vessel holder; f, vessel; g, heat exchanger; h, inlet and outlet tubes; i, heater; j, thermocouple plate; k, rotation shaft. B Batch vessel; and C flow vessel: a, inlet; b, outlet; c, O-ring; d, small compartment; e, baffle plate (reproduced from Kobayashi *et al* 1985 with permission).

reaction vessels, which resemble small conventional fermentors, have a volume of 30 cm^3 and are surrounded by copper-constantan thermopiles. The microbial suspension is stirred with an impeller and air is introduced through a sparger made from sintered steel, pore diameter 50 μm. Standard rates of stirring and air flow are 300 rpm and 15 cm^3 min^{-1}, respectively. As may be expected, the efficient aeration will cause thermal disturbances and the instrument is not suitable where a high sensitivity is required. However, high oxygen consumption is always accompanied by high heat production.

Instead of applying vigorous agitation of the gas-liquid system during aerobic microbial experiments, Dermoun *et al* (1985) used shallow reaction vessels (diameter 42 mm, depth 15 mm) where the medium thus has a large contact area with the gas phase. The vessels, which rest on the thermocouple plates, are equipped with stirrers. Inoculum is injected after the instrument has equilibrated and humidified gas can be introduced to the vessels during the experiments.

Bacteria typically show complex power-time curves when grown on a complex medium and it was first suggested by Boling *et al* (1973) that such fingerprint curves could be used for identification of bacteria. They used an automated 50-channel microcalorimeter for these studies. The instrument, which was an experimental model manufactured by Instrument Laboratory, Lexington, Massachusetts, used simple static ampoules as reaction vessels (Russel *et al* 1975).

Kawabata *et al* (1983) have described a six-channel heat conduction calorimeter for work on soil microbial systems. The calorimetric vessels (volume 30 cm^3), were placed in a hexagonal arrangement with thermocouples interspaced between their bottoms and a large aluminium block used as a common heat sink. A reference calorimetric unit was placed in the centre of the hexagon. The channels were measured stepwise by use of a rotary switch.

So far, there are relatively few examples in biological calorimetry in which Peltier effect cooling is used to keep the reaction vessel at a constant temperature (Hansen *et al* 1980). Figure 3.21 shows schematically the reaction vessel of Tronac's isothermal microcalorimeter (Tronac, Orem, Utah), which is based on the design by Christensen *et al* (1973). A constant Peltier cooling power is balanced against the sum of the heat production from the reaction and the heat pulses released from a control heater. A thermistor is used to control the isothermal conditions. The control heater, thermistor, a stirrer and a titration inlet tube are collected in a unit which can be attached to the vessel. The volume of the reaction vessel can be 2·5–100 cm^3.

Organs and tissues, in particular muscles, nerves and adipose tissues, have been studied extensively by different calorimetric techniques (Woledge 1980, p. 147). In some cases, requirements for electrical and mechanical stimulation and medium perfusion through or around the organs have led to specially designed calorimeters or calorimetric vessels. Several of the instruments used are typical thermopile heat conduction instruments, such as the single vessel calorimeter designed by Lörinzi and Futo (1974), and Lörinzi (1982), which was used in work with muscles. The instrument contained transducers for measurement of mechanical power and displacement. Chinet *et al* (1977) used

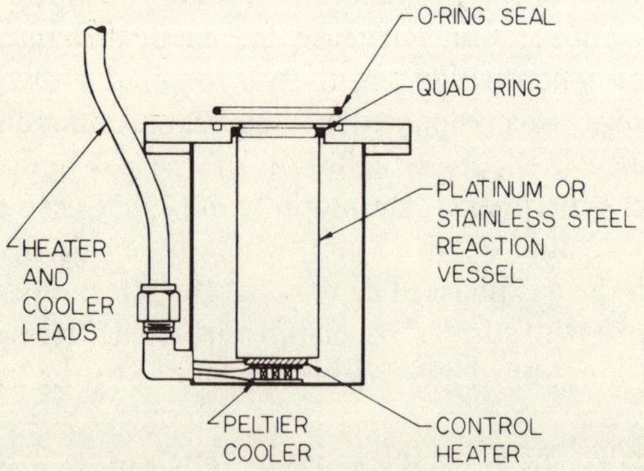

Figure 3.21. Thermoelectric heat pump calorimeter according to Christensen *et al* (1973). An assembly containing control heater, thermistor, stirrer and a titration inlet tube can be attached to the vessel (courtesy of Tronac, Orem, Utah).

a twin perfusion instrument (Secfroid S.A., Lausanne, Switzerland) in work on rat muscle and adipose tissue. Schlipf and Höhne (1981) equipped a twin heat conduction calorimeter (manufactured by BCD Arion, Grenoble, France), with a special vessel for use in muscle stretching experiments. Ponce-Hornos *et al* (1982) used a single vessel design in their work on perfused rabbit interventricular septum (Figure 3.22). The septum was contained in a cavity formed by a circular frame made from Kel F and two platinum-plated copper discs. The thermocouple plates used as sensors were interspaced between the copper discs and the surrounding copper block. The septum, which was perfused via its arterial system, could be stimulated electrically and the tension measured via a transducer.

Dewar vessel calorimeters have been used for measurements of perfused intact hearts (Coulson 1982); however, they were not used as adiabatic calorimeters. Most of the thermal power produced by the organs was carried away by the perfusion fluid and was measured by determination of the temperature difference between the incoming and outgoing flow. Thermocouples were used as temperature sensors. Obviously, this kind of instrument cannot be used when the perfusion rate is zero. Niesler and Axon (1981) used a

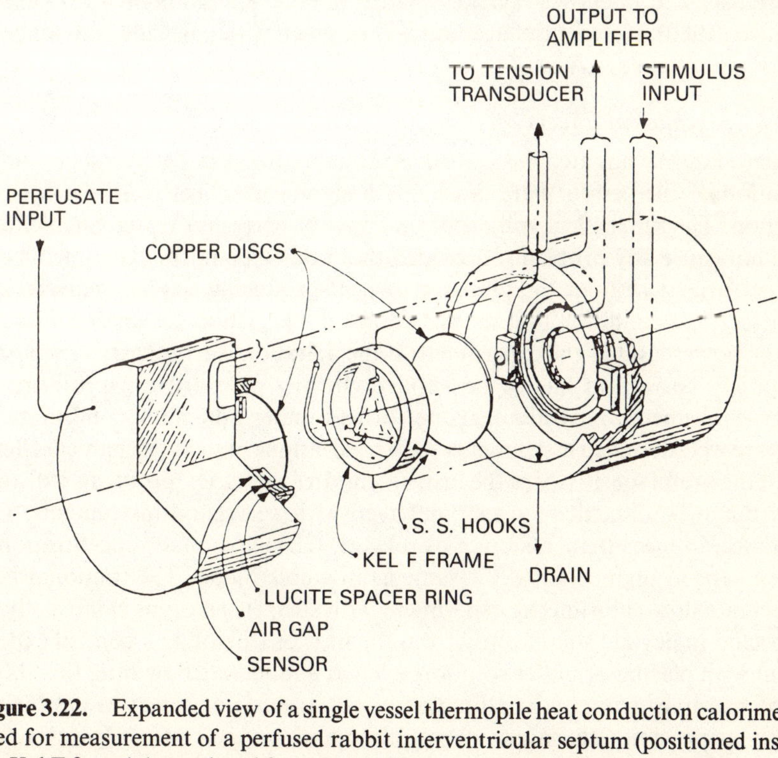

Figure 3.22. Expanded view of a single vessel thermopile heat conduction calorimeter used for measurement of a perfused rabbit interventricular septum (positioned inside the Kel F frame) (reproduced from Ponce-Hornos *et al* 1982 with permission).

similar measurement principle where the thermal detector consisted of a thermopile (a 'Biomedical Seebeck Envelope Calorimeter', Termonetics Corp., San Diego, California). Parameters other than the thermal power (p_{O_2}, aortic and ventricular pressure) were also measured.

3.4 Systematic Errors and Other Problems

Practically all processes are accompanied by heat effects. This makes reaction calorimeters suitable as general process monitors, in addition to their use as thermodynamic instruments. In fact, there is hardly any other physical measurement principle which has equally high resolution on consecutive stages of a complex biological process, or which is as powerful for the detection of unexpected processes or phases of a process. However, it is not easy to analyse, on the molecular level, the different contributions to the sum of processes which give rise to the calorimetric signal from a living system. Another problem is the risk of contributions to the measured signal from many possible sources of systematic errors or from sheer artefacts. Naturally, such problems are particularly important to watch out for when the system investigated has a very low heat production rate. Some of these problems have been discussed in earlier reviews: Wadsö (1975), Spink and Wadsö (1976), Wadsö (1980). Here a brief account will be given with particular reference to work on cellular systems.

3.4.1 Mechanical Effects

When a liquid is agitated in a calorimetric vessel, for example by a stirrer or by rotation of the vessel, there is usually a significant effect from the heat of friction. The same effect will occur in a flow or perfusion vessel. Such effects will not cause any problems, provided that relevant control experiments can be performed without the presence of heat-producing cellular material. Of course, it is a condition that the variations in the friction are small compared to the power level of the experiment. If the agitation is temporary, a separate empirical correction has to be applied. In any case, friction problems in microcalorimetric measurements performed under non-static conditions do require well-designed instruments. Often variations in such mechanical effects limit the useful sensitivity of the instrument. In some cases, problems will arise which can be difficult to overcome, even with very good instruments. For instance, in measurements on erythrocytes, the cell volume concentration is often kept as high as 40%, the same as in whole blood. The frictional heat effect in a flow-calorimetric experiment with such suspensions can usually be expected to deviate significantly from the heat of friction in a control experiment with plasma or buffer solution, even at a moderate flow rate. Possibly a good baseline value can be estimated by comparison of power values for the same suspension determined during flow and stopped flow. However, the physiological conditions may not be quite identical if the cells are moving or if they are under static conditions. Even more difficult situations may arise in flow calorimetry or in stirred batch calorimetric experiments if the cell suspension changes its viscosity significantly during the measurement, e.g. in a bacterial growth experiment. As a general rule, it is wise to keep the frictional power as low as possible (low stirring rate, low flow rate).

In work on living materials, it is very common that the (absolute) power value is determined for the system contained in the reaction vessel. It is then essential that the calorimeter baseline is not only stable, but that its value is also reproducible. (For a thermopile conduction calorimeter, it is rarely exactly zero.) However, it is quite common that significant and unpredictable baseline shifts occur when a vessel is inserted into a calorimeter or when a permanently installed batch vessel is manipulated during cleaning and charging procedures. Figures given for the 'baseline stability' or the instrument 'sensitivity' might be of little interest in such cases. Frequent tests of the reproducibility of the baseline value should therefore be made. In fact, poor baseline reproducibility often limits the usefulness of a calorimeter in quantitative work on cells, but this property is not commonly discussed in reports or in commercial advertisements for calorimetric power meters.

In 'real' microcalorimetry, it is usually not advisable to start a mixing process by breaking a glass ampoule or opening a mechanical valve inside the vessel. The heat effects connected with such operations are generally poorly reproducible.

Measurements carried out at the μW level can sometimes be impaired by heat evolution from physical 'relaxation' processes in the instrument. For instance, small rubber O-rings used to close insertion vessels can give rise to slowly decaying power signals of the order of 1 μW (Suurkuusk and Wadsö 1982). Use of thin teflon packing rings can avoid the effect (Görman Nordmark *et al* 1984).

3.4.2 Evaporation and Condensation

Water is always present in biological experiments and the heat of vaporization of water is very large; 2·4 mJ of heat is absorbed per μg water evaporated at 37 °C. This corresponds to a thermal power of nearly 0·7 μW at an evaporation rate of 1 μg water per h. Clearly, virtually no uncontrolled evaporation or condensation can be tolerated in experiments carried out at the μW level. Possible distillation effects or changes of the gas phase composition must be considered (Spink and Wadsö 1976).

Gas phase problems are avoided if there is no gas phase present, which sometimes gives flow or perfusion calorimeters an extra advantage. Leakage effects due to poor packing rings or pinholes can, of course, be detrimental for a calorimetric experiment, but such effects are usually irregular and will hardly cause any systematic errors.

3.4.3 Adhesion Effects

Many cells adhere easily to the walls of experimental vessels. Sometimes this is intentional, e.g. in work with monolayers of tissue cells. In other cases, adhesion effects can cause serious problems. The adhesion process itself will probably be accompanied by an increased metabolic activity and a significant heat burst. Further, the metabolic activity of attached cells may be different from that of the cells in suspension. In flow experiments, where it is intended that the cells should flow through the calorimetric system, additional problems may occur. The cells can adhere (partially or preferentially) to the flow line before they reach the calorimetric vessel, thereby causing the measured power to be lower than that corresponding to the initial cell concentration. If the cells adhere preferentially to the flow vessel, too high power values will be recorded (Levin 1970, 1973; Wadsö 1980).

The amounts of certain chemical substances, such as hormones and drugs, used in work with cells can be very small, while the wall surface area of a microcalorimetric vessel is quite large. This may lead to significant vessel adsorption effects.

3.4.4 Sedimentation

Partial or complete sedimentation of cells during a calorimetric experiment may lead in some cases to errors and in any case to poorly defined power values. 'Crowding effects' have been reported for some types of cells. In other cases, where one may not talk about 'crowding effects', the metabolic rate can be affected by local depletion of oxygen and differences of pH which develop as a result of the very high cell concentration in the sediment.

In flow calorimetry, heavy cells may sediment in the flow lines or in the flow vessel. This will lead to too low or too high power values, respectively. Measurements with some types of calorimetric vessels may lead to significant errors if local 'hot spots' are formed where the cellular material has been concentrated. For example, with isoperibolic calorimeters where the temperature is measured at one point with a thermistor, it is always important to have an efficient stirring of the calorimetric liquid in order to ensure a uniform temperature.

Problems with sedimentation of cells in flow calorimetry, in particular of yeast cells, have led to the construction of special flow vessels (Gustafsson and Lindman 1977 and Figure 3.11).

3.4.5 Problems with Aerobic Systems

In many types of calorimetric experiments with aerobic systems, it is difficult to arrange for a sufficient supply of oxygen (Wadsö 1980). Water in equilibrium with air dissolves about 0·2 μmol oxygen per cm^3 at 37 °C. 1 cm^3 of aqueous medium will thus contain enough oxygen for the complete oxidation of about 33 nmol glucose. This corresponds to about 10 μW during 3 h for a steady-state process. If the medium is equilibrated with pure oxygen instead of air, the amount of oxygen dissolved will increase by a factor of approximately 5. Thus, with dilute cell suspensions and with very sensitive calorimeters, it is possible to make rather long experiments with aerobic cells without any transfer of oxygen to the calorimetric system during the experiment. In experiments where the liquid medium is shallow and in cases where the cellular material is floating (adipose tissue, fat cells), significant amounts of oxygen can be transferred from a gas phase even if the medium is not agitated.

In flow calorimetry, the oxygen concentration can be monitored conveniently with an electrode positioned in the flow line (Wadsö 1980; Brettel *et al* 1980, 1981; Gnaiger 1983 a, b). It is important that the electrode is positioned as close as possible to the flow cell. Oxygen electrodes have also been placed inside batch calorimetric vessels (Poe *et al* 1967; Nakamura and Matsuoka 1978; Pamatmat 1983), but so far they do not seem to have been used with high-sensitivity calorimeters.

It is fortunate that high rates of oxygen consumption lead to high heat production rates and one may then use comparatively vigorous agitation of the oxygen-liquid system; however, this might damage some types of cells. A different experimental approach is to use vessels with a large contact area between the liquid phase and the gas phase, in combination with moderate agitation, cf. the rotating vessels shown in Figures 3.10, 3.19, and 3.20 and the stirred vessel described by Dermoun *et al* (1985). In some cases, it is suitable to use a mixed (segmented) flow of gas and liquid, for which several flow vessels have been designed (see Figure 3.11). Such vessels are also useful for the prevention of sedimentation of cells.

Two other calorimetric measurement problems with gas-liquid systems should be pointed out. As mentioned above, the enthalpy of vaporization of water is very high. A stream of air perfusing through a calorimetric vessel and not in equilibrium with the calorimetric liquid can thus cause substantial cooling effects. In experiments performed on a high power level, the problem can usually be avoided by a suitable humidifier in series with the flow vessel. In experiments at the μW level, where gas bubbling in addition to oxygenation might be useful as a means of mild agitation, special precautions must be taken (Wadsö 1974, 1980). A segmented flow of gas-liquid, in combination with an adequate heat exchanger, usually provides for a very efficient equilibration of the vapour phase equilibrium.

Another calorimetric problem deals with the fact that the enthalpy of solution of gas into the medium cannot be neglected ($\Delta H = -12{\cdot}0$ kJ mol^{-1} for the solution of oxygen into water at 25 °C). It is therefore often important to know if oxygen consumed by the cells from the solution is replaced by a transfer of oxygen from a gas phase. With efficient agitation of the system, this can probably be assumed to be the case. However, the situation in a static reaction vessel is questionable, in particular if the oxygen consumption rate is high. For this reason it is sometimes preferable to avoid gas phase completely by the use of a flow or perfusion vessel. Note that a corresponding problem can also occur for systems where carbon dioxide is formed.

Special problems can arise with oxygen consumption and carbon dioxide formation in long-time measurements on microbial systems in soil. A calorimetric technique using a permeable silicone rubber membrane has been developed (Ljungholm *et al* 1979; Wadsö 1980).

3.4.6 Calibration and Test Experiments

There is a general and lasting value in accurate and well-documented thermodynamic data for biological systems, However, systematic errors are easily introduced in biological calorimetry, particularly at the microcalorimetric level. An additional source of systematic errors is the calibration (standardization) procedure used for the instrument. On the other hand, with a well performed calibration procedure, it is often possible to cancel out some types of systematic errors.

In good calorimetric work on well-defined chemical or physical systems, meticulous care must be taken to calibrate the instrument properly and to control the experimental procedures. In extreme cases, this can lead to an accuracy of the order of 0·01%. In biological calorimetry, it is of course not meaningful to aim at this kind of accuracy; errors of the order of 1% are usually not significant. However, the errors can easily be much larger with a faulty calibration procedure. Calorimeters are normally calibrated electrically. This is convenient, but it is not always easy to simulate the biological heat effect with an electrical heater in such a way that the calorimetric sensor will react identically.

Normally, reaction vessels have to be designed with many practical factors in mind and it is common that the construction of the calibration heater, and its position in the calorimeter, are far from ideal. If possible, new or modified calorimetric designs should be tested by calibration experiments performed with heaters of different design and in different positions. The experimenter will then get a realistic feeling for the magnitude of possible systematic errors in the calibration value. Difficult calibration problems can arise if flow-through vessels are used in work with cells in suspension (Wadsö 1980).

In order to control electrical calibration values for calorimetric vessels used as power meters, particularly flow vessels, a chemical test or calibration process can be very useful. Recently the power values for several reaction solutions involving the hydrolysis of triacetin in imidazol-acetic acid buffer solutions have been investigated (Chen and Wadsö 1982). Different mixtures giving power values in the range of 7–90 μW cm^{-3} (25, 37 °C) are described. The power values for the mixtures can be calculated accurately ($\pm 0{\cdot}5\%$) as a function of time during extended reaction periods, 20 h or more.

References

Anderson P C and Lovrien R E (1979) *Anal Biochem* **100**, 77–86

Armstrong G T (1964) *J Chem Ed* **41**, 297–307

Bach D and Chapman D (1980) In: Beezer A E (ed) *Biological Microcalorimetry*. London, Academic Press, pp 275–309

Beezer A E, Hunter W H, Lipscombe R P, Newell R D and Storey D E (1982) *Thermochim Acta* **55**, 345–9

Benjamin L (1963) *Can J Chem* **41**, 2210–18

Benoist L and Pithon F (1984) *Thermochim Acta* **72**, 189–94

Boling E A, Blanchard G C and Russel W J (1973) *Nature* **241**, 472–3

Brettel R, Lamprecht I and Schaarschmidt B (1980) *Radiat Environ Biophys* **18**, 301–9

Brettel R, Lamprecht I and Schaarschmidt B (1981) *Eur J Appl Microbiol Biotechnol* **11**, 205–11

Brown H D (ed) (1969) *Biochemical Microcalorimetry*. New York, Academic Press

Calvet E and Prat H (1956) *Microcalorimétrie, applications physicochimiques et biologiques*. Paris, Masson

Calvet E and Prat H (1963) *Recent Progress in Microcalorimetry*. London, Pergamon Press

Chen A-t and Wadsö I (1982) *J Biochem Biophys Meth* **6**, 297–306

Chinet A, Clausen T and Girardier L (1977) *J Physiol* **265**, 43–61

Christensen J J, Gardner J W, Eatough D J, Izatt R M, Watts P J and Hart R M (1973) *Rev Sci Instrum* **44**, 481–4

Coulson R L (1982) *Fed Proc* **41**, 199–203

Danielsson B, Mattiasson B and Mosbach K (1981) *Appl Biochem Bioeng* **3**, 97–141

Dermoun Z, Boussand R, Cotten D and Belaich J-P (1985) *Biotechnol Bioeng* 27, 996–1004

Edsall J T and Gutfreund H (1983) *Biothermodynamics*. New York, Wiley

Eriksson R and Wadsö I (1971) In: Broda E *et al* (eds) *Proc First Europ Biophys Congr Baden* vol 4. Vienna, Wiener Medizinischen Academe, 319–27

Evans W J (1969) In: Brown H D (ed) *Biochemical Microcalorimetry*. New York, Academic Press, 257–73

Fagher B, Monti M and Wadsö I (1985) *Clin Sci* **70**, 63–72

Fortier J-L, Reboul B, Philip P, Simard M-A, Picker P and Jolicoeur C (1980) *J Water Pollut Control Fed* **52**, 89–97

Fujita T, Nunomura K, Kagami I and Nishikawa Y (1976) *J Gen Appl Microbiol* **22**, 43–50

Gnaiger E (1983a) In: Gnaiger E, Forstner H (eds) *Polarographic Oxygen Sensors*. Berlin-Heidelberg, Springer, 134–66

Gnaiger E (1983b) *J Exp Zool* **228**, 471–90

Gonda K, Wada H and Murase N (1985) *Thermochim Acta* **88**, 339–44

Gordon H S and Millero F J (1983) *J Microbiol Meth* **1**, 291–6

Gustafsson L and Lindman B (1977) *FEMS Microbiol Lett* **1**, 227–30

Görman Nordmark M, Laynez J. Schön A, Suurkuusk J and Wadsö I (1984) *J Biochem Biophys Meth* **10**, 198–202

Hammerstedt R H and Lovrien R E (1983) *J Exp Zool* **228**, 459–69

Hansen L D, Jensen T E and Eatough D J (1980) In: Beezer A E (ed) *Biological Microcalorimetry*. London, Academic Press, pp 453–476

Heytler P G and Hardy R W F (1984) *Plant Physiol* **75**, 304–10

Hoffner S E S, Meredith R W J and Kemp R B (1985) *Cytobios* **42**, 71–80

Innskeep P B and Hammerstedt R H (1983) *J Biochem Biophys Meth* **7**, 199–210

Ishikawa Y, Shoda M and Maruyama H (1981) *Biotechnol Bioeng* **23**, 2629–40

Ito E, Sakihama H, Toyama K and Matsui K (1984) *Cancer Res* **44**, 1985–90

Jarrett I G, Clark D G, Filsell O H, Harvey J W and Clark M G (1979) *Biochem J* **180**, 631–8

Kawabata T, Yamano H and Takahashi K (1983) *Agric Biol Chem* **47**, 1281–8

Kemp R B (1975) *Pestic Sci* **6**, 311–25

Kemp R B (1980) In: Beezer A E (ed) *Biological Microcalorimetry*. London, Academic Press, pp 113–30

Kleiber M (1961) *The Fire of Life*. New York, Wiley

Knudsen J. Famme P and Hansen E S (1983) *Comp Biochem Physiol* **74A**, 63–6

Kobayashi K, Fujita T and Hagiwara S (1985) *Thermochim Acta* **88**, 329–38

Kresheck G C (1974) *Cancer Biochem Biophys* **1**, 39–41

Kurtti J J, Brooks M A, Wensman C and Lovrien R E (1979) *J Therm Biol* **4**, 129–36

Kusano K, Nelander B and Wadsö I (1971) *Chem Scripta* **1**, 211–6

Levin K (1970) *Clin Chim Acta* **32**, 87–94

Levin K (1973) *Scand J Clin Lab Invest* **32**, 67–73

Ljungholm K, Norén B, Sköld R and Wadsö I (1979) *Oikos* **33**, 15–23

Lock M A and Ford T E (1983) *Appl Environ Microbiol* **46**, 463–7

Loesberg C, Miltenburg J C and van Wijk R (1982) *J Therm Biol* **7**, 87–90

Lovrien R E, Jorgensson G, Ma M K and Sund W E (1980) *Biotechnol Bioeng* **22**, 1249–69

Lovrien R E, Kurtti T J, Tsang R and Brooks-Wallace M (1981) *J Biochem Biophys Meth* **5**, 307–18

Lörinzi D and Futo Z (1974) *Acta Biochem Biophys Acad Sci Hung* **9**, 371–81

Lörinzi D (1982) *Studia Biophys* **89**, 81–97

Martin C J and Marini M A (1979) *CRC Crit Rev Anal Chemistry* **8:3**, 221–83

Mattiasson B, Larson P-O and Mosbach K (1977) *Nature* **268**, 519–20

McKinnon I R, Fall L, Parody-Morales A and Gill S J (1984) *Anal Biochem* **139**, 134–9

Monk P and Wadsö I (1968) *Acta Chem Scand* **22**, 1842–52

Monti M and Wadsö I (1976) *Scand J Clin Lab Invest* **36**, 565–72

Mou D G and Cooney C L (1976) *Biotechnol Bioeng* **18**, 1371–92

Nakamura T and Matsuoka I (1978) *J Biochem* **84**, 39–46

Niesler R A and Axon D W (1981) *Phys Med Biol* **26**, 1069–85

Nunomura K and Fujita K (1982) *J Gen Appl Microbiol* **28**, 479–90

Pamatmat M M (1983) In: Gnaiger E, Forstner H (eds) *Polarographic Oxygen Sensors*. Berlin-Heidelberg, Springer, pp 167–75

Picker P, Leduc P-A, Philip P R and Desnoyers J (1971) *J Chem Thermodynamics* **3**, 631–42

Poe M, Gutfreund H and Eastabrook R W (1967) *Acta Biochem Biophys* **122**, 204–11

Ponce-Hornos J E, Ricchiati N V and Langer G A (1982) *Am J Physiol* **243**, H289–H295

Privalov P L (1980) In: Beezer A E (ed) *Biological Microcalorimetry*. London, Academic Press, pp 413–51

Prosen E J (1973) *National Bureau of Standards Report NBSIR*-73-179. Washington, Nat'l Bureau Stds

Randzio S and Suurkuusk J (1980) In: Beezer A E (ed) *Biological Microcalorimetry*. London, Adacemic Press, pp 311–41

Russel W J, Zettler J F, Blanchard G C and Boling E A (1975) In: Heden C-G, Illeni T (eds) *New Approaches to the Identification of Microorganisms*. New York, Wiley, pp 101–21

Schaarschmidt B and Reichert U (1981) *Exp Cell Res* **131**, 480–3

Schlipf M and Höhne G W H (1981) *Thermochim Acta* **49**, 1–9

Spink C H (1980) *CRC Crit Rev Anal Chemistry* **9:1**, 1–95

Spink C and Wadsö I (1976) In: Glick D (ed) *Methods of Biochemical Analysis* Vol. 23. New York, Wiley, pp 1–158

Sturtevant J M (1971) In: Weissberger A, Rossiter B W (eds) *Physical Methods of Chemistry* Vol. 1, Part V. New York, Wiley, pp 347–425

Suurkuusk J and Wadsö I (1982) *Chemica Scripta* **20**, 155–63

Takahashi K (1973) *Agric Biol Chem* **37**, 2743–7

Wadsö I (1968) *Acta Chem Scand* **22**, 927–37

Wadsö I (1970) *Q Rev Biophys* **3**, 383–427

Wadsö I (1974) *Science Tools* **21**, 18–21

Wadsö I (1975) In: Pain R, Smith B (eds) *New Techniques in Biophysics and Cell Biology* Vol. 2. New York, Wiley, pp 85–126

Wadsö I (1980) In:Beezer A E (ed) *Biological Microcalorimetry*. London, Academic Press, pp 247–74

Woledge R C (1980) In: Beezer A E (ed) *Biological Microcalorimetry*. London, Academic Press, pp 145–62

Chapter 4

Growth and Metabolism of Bacteria

A. M. James

4.1 Introduction

A prominent feature of microbial growth processes is the generation of large amounts of heat. Although the removal of heat generated in large commercial fermentors is a well known problem, the cooling requirements are often calculated by some known or proposed material balance rather than measured directly by calorimetric methods (Swartz 1985). In addition to providing information on such problems, modern flow calorimeters provide an alternative non-destructive method of monitoring changes in microbial metabolism. The measured waste heat represents the algebraic sum of heat from all the complex metabolic processes, rather than the heat of some isolated process. Thus calorimetry is non-specific and thermal data must be supplemented by other physical, chemical or biological data to permit meaningful interpretation.

The simplest case of microbial metabolism is that of catabolism uncoupled from anabolism. Washed bacteria suspended in buffered glucose solution establish a constant rate of glycolysis and therefore a constant rate of heat production (power). No work is done by the cells in increasing biomass or accumulating reserve material; a small amount of energy is expended in translocation of glucose into the cytoplasm and in excluding or accumulating ions. Although equilibrium thermodynamics cannot be applied to such an open system, the calculated molar enthalpy change for the degradation of glucose by bacteria and erythrocytes is in agreement with the measured heat production.

In a nutrient medium with glucose as the sole carbon and energy source where anabolic and catabolic processes are closely coupled, the bacteria grow converting glucose into cellular material and carbon dioxide with the release of energy. Growth is accomplished as a result of the bacteria performing work in transporting nutrients and synthesizing and organizing cellular structures. The energy for these processes is obtained from the catabolic processes involved in the breakdown of glucose; some of the energy is conserved in the ATP pool. The heat liberated is the balance between the enthalpy of anabolic and catabolic processes.

With the improvements in instrumentation and techniques for monitoring other growth parameters there has been an increase in the applications of calorimetry in the biological sciences; many books and proceedings of conferences have been published (Wadsö 1970, 1975, 1980, 1983; Forrest 1972; Zotin and Lamprecht 1976; Lamprecht and Schaarschmidt 1977; Beezer 1977, 1980).

Recently there has been a change in the emphasis from qualitative studies describing the shape and position of power-time traces during bacterial growth to more quantitative studies leading to the determination of enthalpy changes. This has been made possible by the improvements in instrumentation and the realization of the importance of rigorous control of growth and test conditions and the need for the simultaneous monitoring of other growth parameters. This is a review of more recent work on the applications of microcalorimetry to the study of the kinetics and thermodynamics of normal and inhibited bacterial growth.

4.2 Estimation of the Enthalpy of Metabolism

The energy required to drive the biosynthetic or anabolic processes in a cell is supplied by the exergonic catabolic processes. Thermal studies of bacterial growth only provide data on the rate of production of, and the total amount of, waste heat; as yet it is not possible to attribute particular amounts of energy to definite metabolic processes.

The heat evolved, q or $\Delta_{\text{met}}H$, is the balance between the enthalpy of the anabolic and catabolic processes (Belaich 1980);

$$q = \Delta_{\text{met}}H = q_{\text{cat}} + q_{\text{an}} = (1-\alpha)\Delta_{\text{cat}}H + \alpha\Delta_{\text{an}}H \tag{4.1}$$

where α represents the fraction of the carbon source incorporated into the cellular material and can be calculated:

$$\alpha = \frac{y \times Y_s}{100 \times 12 \times a} \tag{4.2}$$

where y is the weight (g) of carbon in 100 g cells (dry weight), Y_s is the molar growth yield (g mol^{-1}), 12 the atomic mass of carbon and a the number of carbon atoms in 1 mol of energy and carbon source. A more accurate value for

α may be obtained from a knowledge of the experimental distribution of carbon between biomass and carbon dioxide.

When the nitrogen source is simple (NH_3, NO_3^-) the enthalpy change corresponding to anabolism can be estimated from a knowledge of the elemental composition of the biomass and the molar growth yield. When cells are grown in defined medium under substrate-limiting conditions, values for $\Delta_{cat}H$ and $\Delta_{an}H$ can be calculated provided that the products of metabolism and the stoichiometries are well established.

The enthalpy of formation of cells, $\Delta_f H$, can be obtained from a knowledge of the empirical formula and the enthalpy of combustion, $\Delta_c H$, of the biomass. For the combustion of glucose-grown cells of *Klebsiella aerogenes* ($M_r = 24{\cdot}98$):

$$CH_{1{\cdot}77}O_{0{\cdot}48}N_{0{\cdot}25} + 1{\cdot}20O_2 \rightarrow CO_2(g) + 0{\cdot}885H_2O(l) + 0{\cdot}125N_2(g)$$

$\Delta_c H = -22{\cdot}7\ \text{kJ g}^{-1}$, whence $\Delta_f H$ (cells) $= -79{\cdot}5\ \text{kJ mol}^{-1}$. The enthalpy change ΔH for the synthesis of 1 mol of cells from glucose and ammonia:

$$1/6(C_6H_{12}O_6) + 0{\cdot}25NH_4OH \rightarrow CH_{1{\cdot}77}O_{0{\cdot}48}N_{0{\cdot}25} + 0{\cdot}74H_2O + 0{\cdot}015O_2$$

calculated from the enthalpies of formation of the reactants and products is $\Delta H = 10{\cdot}94\ \text{kJ mol}^{-1}$ (i.e. an endothermic process). $\Delta_{an}H$ (Equation 4.1) is the enthalpy change corresponding to the synthesis of cells when 1 mol of substrate (glucose) is fully transformed into cellular material, i.e. $\Delta_{an}H = 6 \times 10{\cdot}94 = 65{\cdot}94$ kJ (mol glucose)$^{-1}$. The molar growth yield of *K.aerogenes* is 89·3 g (mol glucose)$^{-1}$, hence $\alpha = 0{\cdot}596$ (Equation 4.2). Thus $q_{an} = \alpha\Delta_{an}H = 39{\cdot}1\ \text{kJ mol}^{-1}$. The quantity of heat associated with the fraction $(1-\alpha)$, i.e. 0·404, of glucose during the metabolism of 1 mol of glucose, $q_{cat} = -1136$ kJ (mol glucose)$^{-1}$. Thus the enthalpy change on metabolism of 1 mol of glucose, $\Delta_{met}H = -1136 + 39{\cdot}1 = -1097$ kJ.

From the calculated and experimental values of $\Delta_{met}H$ Traore *et al* (1981) were able to differentiate between alternative metabolic processes. This method of calculation is only possible when the carbon source, acting as both energy and carbon source, is completely consumed during the growth of the organisms; it is thus essential to establish a carbon balance.

The accuracy of $\Delta_{met}H$ (calc) is very dependent on the values of y and Y_s; 2% errors in either of these experimentally determined values can give rise to a 10% error in $\Delta_{met}H$.

4.3 Experimental Arrangements for Monitoring Heat Production

4.3.1 Microcalorimeters, Vessel Design

Several different calorimetric principles and a wide variety of practical designs have been used for the study of the thermal properties of biological systems

(p. 34). Thermopile heat-conduction type calorimeters are most suitable for the study of processes in which there are small heat effects and for processes of long duration. Most modern calorimeters for biological work, based on this principle, are arranged as twin instruments in which signals from two calorimetric units are connected in opposition. It is thus a differential signal which is recorded. The main advantage of this is that thermal disturbances from the surroundings, which influence both vessels to the same extent, cancel out. In addition to the microcalorimeters available commercially, various heat conduction calorimeters have been designed for specific purposes (Lovrien *et al* 1980; Ishikawa *et al* 1981a; Lock and Ford 1983).

The reaction vessel in the calorimeter may be either a batch or a flow cell. In the simplest type of batch cell (volume 1–15 cm^3) growth occurs under changing environmental conditions; there is no means for stirring or aeration of the medium, sampling the culture or controlling the composition of the gas space above the liquid. A range of different vessels have been described (Wadsö 1980 and p. 48) in which the growth medium can be agitated by a piston (Lamprecht and Schaarschmidt 1973), stirring and aerated (Dermoun *et al* 1985) or irradiated by a fibre-optic light guide (Schaarschmidt and Lamprecht 1973; McIlvaine and Langerman 1977).

Alternatively the culture can be grown in a fermentor outside the calorimeter and the culture pumped through a flow cell in good thermal contact with the heat sink. Stainless steel cylindrical flow cells (volume ca. 1 cm^3) have been designed (Figure 3.11) to overcome thermal disturbances caused by air bubbles (Eriksson and Wadsö 1971; Gustafsson and Lindmann 1977). The culture, after passage through the calorimeter, may be recycled to the fermentor (Nichols 1980; James and Djavan 1980) or pumped to waste (Semenitz 1980; Leiseifer and Schleser 1983). Careful temperature control of the fermentor, calorimeter and connecting tubing is essential to avoid thermal disturbances.

Growth in an outside fermentor has the advantages that aeration conditions can be controlled and samples for analysis removed from, or material (e.g. substrate, antibacterials) added to, the growing culture without causing any thermal disturbance. Organisms growing in a batch cell within the calorimeter are subject to changes in the gas environment and different gas levels in the medium, and sedimentation (p. 62) which can, in turn, give rise to changed metabolism. Determination of biomass and other properties of the cells or medium in parallel cultures in a similar vessel and under similar environmental conditions outside the calorimeter is a poor substitute for measurements on the actual culture in the calorimeter (Dermoun *et al* 1984). In the flow-cell technique care must be taken to ensure that wall growth in the cell and connecting tubing (p. 62) does not give false thermal data.

A typical fermentor-flow-calorimeter assembly, used by a large number of workers, with the facility to monitor a range of other biological and physical properties is shown in Figure 4.1. Growth in the fermentor may be either

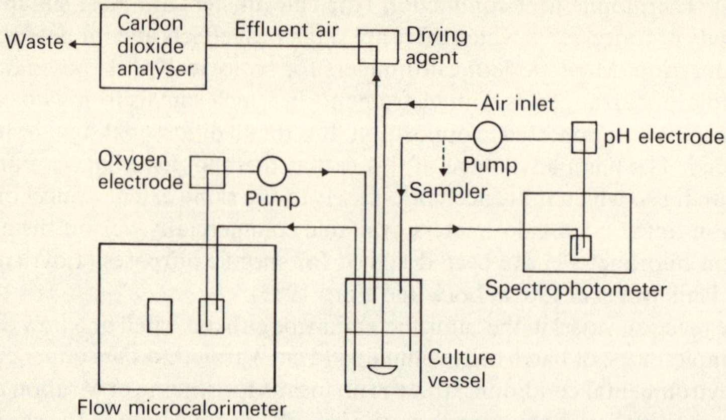

Figure 4.1. Schematic diagram of flow-microcalorimeter coupled to a fermentor for monitoring changes of thermal and growth parameters during bacterial growth.

batch-wise (Glanser *et al* 1979; Nichols *et al* 1979) in which the cells are grown through one complete growth cycle to a final stationary population or in a chemostat (James and Djavan 1980, 1981, 1982; Leiseifer and Schleser 1983) in which the steady state population is controlled by substrate limitation.

As an alternative to heat conduction calorimeters, a special temperature control system has been developed and applied to the continuous monitoring of the heat evolved during growth in a fermentor (Cooney *et al* 1968; Luong and Volesky 1980, 1982b). The fermentation broth is 'overcooled' by a defined constant cooling water flow; any excess heat removed from the fermentor is made up by an electric heater. Prior to inoculation of the growth medium, the heat production by the immersion heater remains constant; this is recorded as a constant voltage across the heater. During exponential growth and increase in heat output there is a marked decrease in the voltage; with the onset of the stationary phase and the cessation of microbial activity the voltage increases.

A non-isothermal calorimeter, operating directly in the fermentor, aimed at measuring heat flow in aerobic microbial growth processes with high biomass production and high oxygen transfer rates has been described (Heinritz *et al* 1983). The use of *in situ* calorimetry is advantageous because it has a low response time and is not limited by high cell concentrations or the presence of non-cellular solids. A potential problem is associated with changes in the fluid properties (e.g. viscosity) which can alter the power drain in the fermentor (Mou and Cooney 1976).

The output from any of these thermal monitoring devices is in the form of a power-time, *p-t*, trace (Figure 4.2), formerly known as a 'thermogram'.

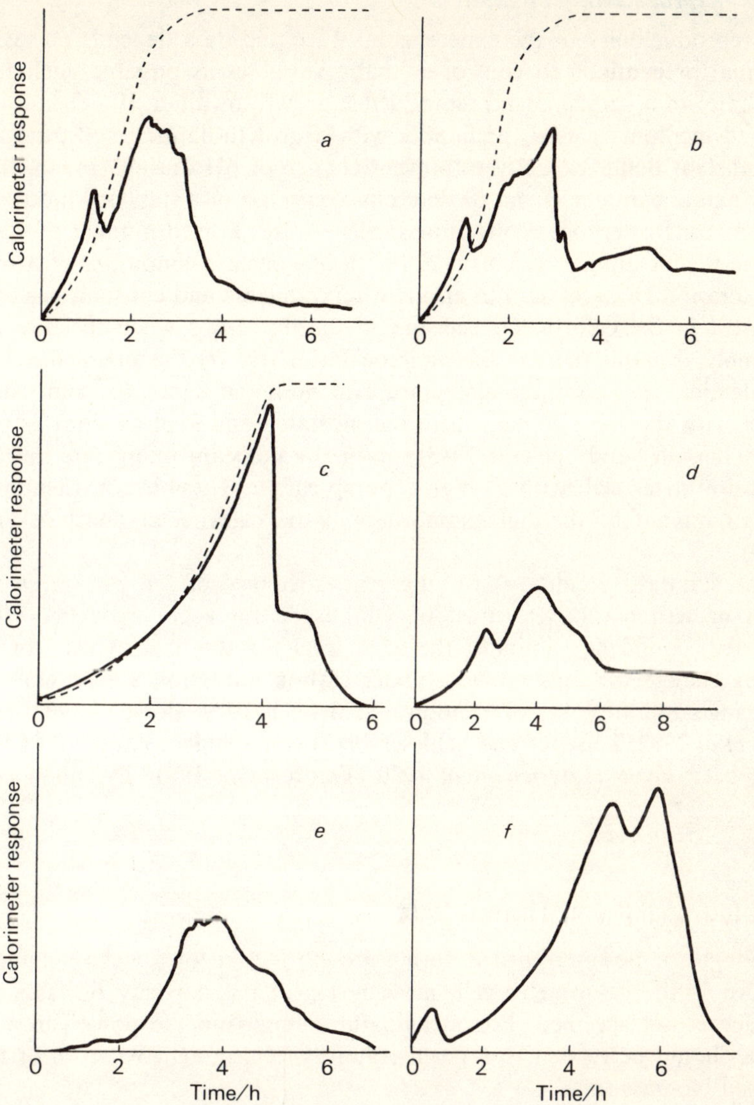

Figure 4.2. Power-time traces for different organisms. (- - - - - growth curve). *a E.coli* 392 growing in nutrient medium, flow calorimetry (after Semenitz *et al* 1982). *b Serratia* 104 growing in nutrient medium, flow calorimetry (after Semenitz *et al* 1982). *c K.aerogenes* NCTC 418 growing in glucose-limited medium with poor aeration, flow calorimetry (after Bowden 1983). *d S.aureus* H 13665 growing in Columbia broth, flow calorimetry (after Semenitz 1979a). *e Enterobacter cloacae* and *f P.aeruginosa* growing in meat extract broth with added peptone, glucose, vitamins, ampoule calorimetry (after Kallings and Hoffner 1977).

4.3.2 Reproducibility of Data

The reproducibility of the experimental trace and data depends on taking essential precautions to control as many variables as possible, including; (*a*) choice of a standard and reproducible growth medium. If a chemically-defined medium is used, preferably with a growth-limiting substrate and of sufficient buffer capacity to prevent change of pH during growth, thermodynamic parameters can be determined; (*b*) use of a standard inoculum. An ampoule of cryogenically stored cells ensures a fixed number of viable cells in the inoculum (Nichols *et al* 1979); (*c*) constant conditions of stirring and aeration (if necessary) to ensure a reproducible and constant gas environment in the calorimeter cell and in the fermentor when the flow cell assembly is used; (*d*) for flow microcalorimetry; (*i*) the maintenance of an identical and constant temperature in both the fermentor and calorimeter with the tubes joining them thermostated and kept as short as possible (Leiseifer and Schleser 1983); and (*ii*) a constant pump rate through the calorimeter cell, with as high a pump rate as possible consistent with the attainment of thermal equilibrium in the calorimeter (Nichols *et al* 1979).

The standard conditions for the reproducibility of the *p-t* traces and other growth parameters must be established for each organism. When reported, confidence limits at the 95% level for the heat evolved or enthalpy change for cells growing under carbon limitation are generally in the range ± 1 to $\pm 4\%$ (Dermoun and Belaich 1979; Bowden 1983; Herman *et al* 1982; Leiseifer and Schleser 1983); some higher limits up to 16% have been given (Lovrien *et al* 1980; Traore *et al* 1981; Dermoun *et al* 1984).

4.4 Interpretation of Thermal Data

Although the power recorded during growth is that for a given volume of culture in the calorimeter cell, nevertheless in the majority of cases this volume is not specified. For meaningful comparison the power, in a *p-t* trace, should be related to a given volume of culture ($mW\ cm^{-3}$) or to a defined biomass ($mW\ g^{-1}$).

The total heat produced by the organisms is obtained from the integration of the *p-t* trace; during the growth phase in either defined or complex nutrient media the heat produced increases exponentially.

From the thermal and other data (biomass, carbon dioxide, substrate concentrations) collected during the growth of cells in defined medium enthalpy changes associated with metabolism, $\Delta_{met}H$, and with the production of biomass (thermal yield) $\Delta_p H$, can be calculated and mass and energy balances established.

4.4.1 Thermal Data from Cells Growing in Defined Medium with Growth-limiting Substrate(s)

The simplest *p-t* trace is obtained for cells growing in carbon-limited defined medium in which the carbon substrate is the sole carbon and energy source (Figure 4.3). During the logarithmic growth phase the power increases exponentially; when the substrate is completely exhausted and growth ceases the power decreases rapidly to the base line (Belaich 1963; Bovinet 1964; Schaarschmidt and Lamprecht 1976; Dermoun and Belaich 1979; Nichols and James 1980; Traore *et al* 1981; Herman *et al* 1982; Leiseifer and Schleser 1983; Dermoun *et al* 1985). The sudden decrease of power is due to the exhaustion of the carbon and energy source and not to the accumulation of toxic products (Belaich *et al* 1968).

The carbon dioxide-time profile during the aerobic growth of *K.aerogenes* in glucose-limited medium is parallel to the *p-t* trace (Nichols 1980; Bowden and James, 1985); the first order rate constants for the production of biomass, power and carbon dioxide are equal. No secondary intermediates are formed under these growth conditions and all the carbon from the substrate may be apportioned between biomass and carbon dioxide.

The *p-t* profile and derived thermal parameters for the growth of glucose-adapted cells of *K.aerogenes* in glucose-limited medium containing added α-methyl-D-glucoside (αMG) are markedly different from those obtained during growth in glucose alone (Nichols and James 1983). The rate constants

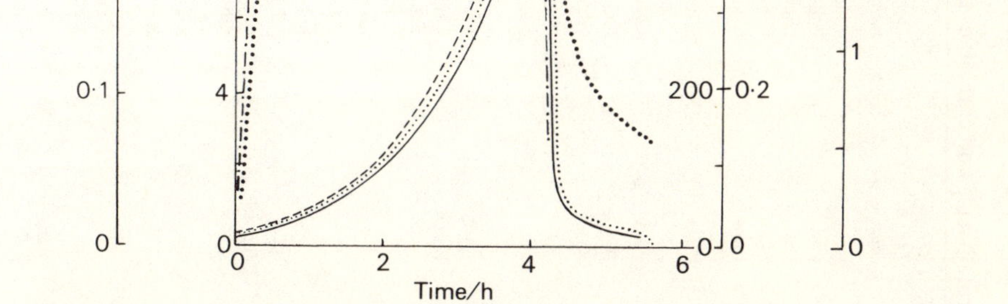

Figure 4.3. Power-time, specific power-time, ATP and carbon dioxide profiles for the growth of *K.aerogenes* (NCTC 418) in glucose-limited medium (after Bowden and James 1985a). ----- biomass; ——— power; —·—·— specific power; ········ ATP; •••••••• carbon dioxide.

Table 4.1 Thermal yields and enthalpies of metabolism for a range of organisms

Organism	Type of calorimeter	Temp.	Limiting substrate	Thermal yield $-\Delta_p H$/kJ g^{-1}	$-\Delta_{met}H$/kJ mol^{-1} Anaerobic	Aerobic	Theoretical	Comments	Reference
Klebsiella aerogenes	Flow	37	Glucose	9·5					Nichols *et al* (1979)
NCTC 418	Flow	37	Glucose	13·5		1069			Nichols and James (1980a)
	Flow	25–40	Glucose	13·8		1130		Variation with temp.	Nichols and James (1981a)
	Flow	42	Glucose	40·0		1800			
	Flow	37	Glucose	9·5		1277	1100		Bowden and James (1985a)
	Flow	37	Glucose	17·0		1340	1292	Phosphate limitation	Bowden and James (1985b)
	Flow	37	α-methyl-D-glucoside	29·0–19·0		2960–1750		Values depend on conc. of αMG	Nichols and James (1983)
	Flow	37	Glucose	9·4		1128	1086		Bunker (1985)
Klebsiella aerogenes	Batch	37	Glucose	3·9*	117				Bovinet (1964)
	Batch	37	Glucose	11·0					Grangetto (1963)
Escherichia coli		37	Glucose	13·7				Medium enriched with amino acids	Cooney (1968)
		37	Glucose	19·8					
Escherichia coli	Flow		Glucose			818			Ericksson and Holme (1973)
Escherichia coli K.12.322	Batch	30	Glucose	4·0*	123				Belaich and Belaich (1976)
			Fructose	4·5*	123				
			Galactose	4·0*	106				
			Lactose	4·1*	223				
E.coli 376			Glucose	3·7*	105				
			Galactose	4·3*	117				
			Lactose	4·0*	226				
E.coli B/5	Batch	25	Glucose		142	1272		Limiting values as the concentration of the substrate approaches zero	Lovrien *et al* (1980)
			Glycerol		42	841			
			Fructose		134	1343			
			Mannitol		84	1456			
			Sorbitol		192	1598			
			Mannose		105	1025			
			Galactose		188	962			
			Xylose		38	745			
E.coli	Batch	30	Glucose			1075		Continuous culture value depends on D	Ishikawa and Shoda (1983)
			Glucose			460–956			

E.coli		37	Glucose	12·8				Monitoring in culture	Luong and Volesky (1982a)
E.coli K.12.322	Flow	30	Succinate	18·5		733	844		Herman *et al* (1982)
	Batch	30	Succinate	19·6		739	785		Dermoun and Belaich (1979, 1980)
E.coli K.12	Flow	37	Glucose	21·0		1397			Leiseifer and Schleser (1983)
E.coli K.12.322	Batch	30	Glucose	16·6		1257	1412		Dermoun *et al* (1985)
			Acetate	35·5		476	618		
			Succinate	19·7		739	792		
Zymomonas mobilis	Batch	37	Glucose		133		135		Belaich *et al* (1968)
Strepococcus lactis	Batch	37	Glucose	6·6*	122				Bovinet (1964)
Acinetobacter B-1	Batch	37	Benzene			800			Lovrien *et al* (1980)
			Toluene			790			
Pseudomonas putida	Batch	25	Naphthalene			3390			
			2-methyl-naphthalene			2730			
Bacillus stearothermophilus	Batch	47–62·5	Glucose	30–35*	900–1620			Values depend on nitrate concentration and temp.	Reiling and Zuber (1983)
Vibrio alginolyticus	Batch	25	Glucose			1190			Gordon and Millero (1980)
	Flow	25	Glucose			1274			
	Batch/Flow	25	Glucose			1340			Gordon *et al* (1982)
Desulphovibrio vulgaris (NCIB 8303)	Batch	30	Lactate		36		30		Traore *et al* (1981)
			Pyruvate		70		79		
Cellulomonas uda, sp.21399	Batch	30	Glucose	8·2		1079	1314		Dermoun *et al* (1984)
			Cellobiose	6·3		935		Values calculated on basis of glucose equivalent	
			Cellulose	9·85		1130			
			Cellulose	13·3		6**		**kJ g^{-1}	Dermoun and Belaich (1985)

Table 4.1 (contd)

Organism	Type of calorimeter	Temp.	Limiting substrate	Thermal yield $-\Delta_p H$/kJ g^{-1}	$-\Delta_{met}H$ kJ mol^{-1} Anaerobic	Aerobic	Theoretical	Comments	Reference
Soil organisms (mixture)	Batch	30–48	Range of sugars		14–15**			**kJ g^{-1}	Yamano and Takahashi (1983)
Candida utilis		37	Glucose	12·1		1046		Monitoring in culture	Luong and Volesky (1982a)
			Sucrose	8·8					
			Ethanol	28·9					
Candida lipolytica			Glucose	1·0		1046			
			Hexadecane	28·8					
Candida intermedia			Glucose	12·6					
Organisms grown in nutrient medium with no growth-limiting substrate									
K.aerogenes NCTC 418	Flow	37	Nutrient broth	1·6					Bunker (1985)
			Glucose + nutrient broth	9·0				Glucose-limited medium supplemented with 10% Oxoid nutrient broth	
Staph.aureus 13137	Flow	37	Nutrient broth	3·8					
Ps.47			Nutrient broth	6·5					
Pseudomonas aeruginosa	Flow	37	Nutrient broth	2·6					

*Thermal yield under anaerobic conditions.

**Only mean values quoted—at 95% level error limits may be as much as ±12%.

for biomass and power production decrease linearly with increasing concentration of αMG. Yield and enthalpy data indicate that αMG is not metabolized; this is not supported by the carbon dioxide production. The *p-t* traces for αMG-adapted cells are basically similar to those for glucose-adapted cells, although the maximum power occurs before the end of exponential growth. The values of $\Delta_p H$ and $\Delta_{met} H$ for αMG-adapted cells growing in medium with αMG as sole carbon and energy source, which depend on the concentration of αMG, are always less than the corresponding values for glucose-adapted cells (Table 4.1). The efficiency of αMG-adapted cells for glucose is greater than that of glucose-adapted cells, indicating that a second transport system has been developed which is not only specific to αMG.

Cells of *K.aerogenes* adapted to potassium limitation, with excess glucose present, exhibited similar thermal properties as for growth under glucose limitation until the potassium in the medium was exhausted; the remaining glucose was then slowly fermented with the dissipation of much heat (Nichols 1980).

The cumulative heat released can be used as a physiological parameter to assess the growth of microorganisms (Luong and Volesky 1982a). Although the quantity of heat released per dry weight of biomass (thermal yield) depends on the substrate and molar growth yield, nevertheless for four different organisms (including yeasts) growing on glucose the value of the thermal yield (12·12 kJ g^{-1}) is not dependent to any significant degree on the type of organism. For the fermentation of *n*-alkanes thermal yields are in the range 18·97–32·66 kJ g^{-1} (Cooney 1968). The correlation between the cumulative heat release and the total cell population can be used to calculate the heat of combustion of microbial cells (Swartz 1985).

The rate of oxygen uptake (OUR) also follows the trend of variation in the power output; the OUR, like the power, declines drastically with the onset of the stationary phase (Luong and Volesky 1980). When *E.coli* is grown in a 1:1 mixture of glucose and lactose the heat and oxygen rate curves exhibit dual maxima, the first peak occurring when glucose (the preferred carbon source) is exhausted and the second on the exhaustion of both sugars and the end of growth. Linear relationships were established between the power and the OUR and between the total heat released and the total oxygen consumed. For a range of organisms (including yeasts) the total heat released from a range of carbon substrates is given by:

$$q = 0{\cdot}461 \pm 0{\cdot}084 \text{ MJ (mol } O_2)^{-1}$$

The experimental results agree with the theoretical predictions:

$$q = 0{\cdot}444 \pm 0{\cdot}017 \text{ MJ (mol } O_2)^{-1}$$

(Minkevich and Eroshin 1973). Cooney *et al* (1968) report a proportionality constant of 0·460 ± 0·042. The correlation between oxygen consumption and total heat released has considerable potential for predicting heat evolution from the metabolism of microorganisms on simple or complex substrates.

Under conditions of partial oxygen deficiency, either in the calorimeter cell or in the fermentor, or during anaerobic growth, the power does not return sharply to the baseline on the exhaustion of substrate (Figure 4.2c) but has a shoulder or more prolonged period of constant lower power output (Eriksson and Wadsö 1971; Nichols *et al* 1979; Nichols and James 1980; Ishikawa *et al* 1981b; Gordon *et al* 1982; Dermoun *et al* 1984, 1985). This secondary low power output is due to the exothermic breakdown of secondary intermediates (e.g. acetate) which have accumulated during exponential growth. The carbon dioxide-time trace again parallels the *p-t* trace, indicating that the exothermic metabolic processes result in the production of carbon dioxide without the formation of biomass.

To compare the efficiency of energy utilization at different stages of growth or under different conditions of growth, the specific power, *P*, defined as the power per unit dry weight of cells ($W\ g^{-1}$), is calculated by dividing the instantaneous power by the biomass at the same time. The specific power increases rapidly during the initial stages of the exponential growth of *K.aerogenes* in glucose-limited medium to a high value, thereafter decreasing to a lower constant value, with small oscillations about a mean value, during the remainder of growth (Figure 4.3). At the onset of the stationary phase the specific power decreases to zero. The high values recorded in the early stages of growth indicate that this is the most energetically inefficient period, in support of the observations of Schaarschmidt *et al* (1975) and Gustafsson (1979), and the interpretation of Bayne-Jones and Rhees (1929) that 'young cells produced more heat than old ones'. These high values may be due in part to errors caused by the small number of organisms present and in part to an increased relative rate of catabolism (Poole and Haddock 1975), probably the result of ATP-requiring exothermic reactions activated to repair and/or to replace cellular components of the cells comprising the inoculum.

The variation of the size of the ATP pool of cells of *K.aerogenes* grown under conditions of glucose limitation (Figure 4.3) follows that for the specific power (Bowden and James 1985a), suggesting that both parameters are indicators of similar metabolic processes. The specific power is more dependent on the rate of catabolism rather than on the degree of coupling, while the reverse is true for the ATP pool. During the stationary phase metabolic processes for the maintenance of cellular integrity occur; the small but measureable size of the ATP pool is determined partly by these requirements. On the other hand, the actual maintenance energy is too small to be measured and hence the power and specific power values are zero.

The stability of the size of the ATP pool and its reluctance to fall below a certain value in resting cells is remarkable; presumably when all the carbon source has been exhausted the supply of ATP and demands for its use in biosynthetic processes cease simultaneously. It appears that there is a mechanism whereby the ATP pool is given priority with the obvious advantage to the cell of being available and able to respond suddenly to more favourable growth conditions after a period of deprivation.

For the establishment of a complete energy balance the sum of the energy stored in biomass (and other intermediates) and that wasted must be compared with the total available energy in the initial carbon substrate. Payne (1970) reported that 55–65% of the available energy was converted to biomass irrespective of the substrate or the organism. For *K.aerogenes* grown aerobically in 3·3 mmol dm^{-3} glucose the proportion of energy stored in biomass is 55·4% and that wasted is 32·3%; i.e. 87·7% recovery of the available energy. The balance, i.e. 12·3%, is believed to be composite in nature, part being used for biosynthetic and maintenance processes while part is stored in macromolecular compounds within the cell (Bowden and James 1985a). While the proportion of the total energy stored in the cells remains constant with increasing glucose concentration, there is a decrease in the waste energy and a corresponding increase in the residual energy. For *K.aerogenes* the stored and waste energy are constant at temperatures below 40 °C; at higher temperatures there is a dramatic decrease in the stored energy and increase in the waste energy (Nichols and James 1981a). For cells of *Escherichia coli* growing on succinate the residual energy is 9% (Dermoun and Belaich 1979) and for *Bacillus stearothermophilus* growing on glucose 10–29% depending on the nitrate concentration of the medium and the temperature of growth (Reiling and Zuber 1983).

For *Desulphovibrio vulgaris* growing anaerobically in sulphate medium containing either lactate or pyruvate as the sole growth limiting factor, the *p-t* trace is simple (compare Figure 4.3); the end products include acetate, ethanol, carbon dioxide, hydrogen suphide and molecular hydrogen (Traore *et al* 1981). The experimental and theoretical values of $\Delta_{met}H$ for both substrates were used to confirm the mechanism by which the substrates are metabolized. The production of hydrogen constitutes an efficient regulating system of electrons, from the energy source through the pathway of sulphate reduction.

In mixed cultures of *D.vulgaris* and *Methanosarcina barkeri* with lactate or pyruvate as the carbon and energy source, two power pulses were observed (Traore *et al* 1983); the first described the growth of *D.vulgaris* utilizing trace amounts of sulphate as the electron acceptor. After the exhaustion of sulphate the power decreased dramatically; the second peak described the growth of the mixed culture with *M.barkeri* acting as the hydrogen acceptor. The baseline was reestablished on the exhaustion of the carbon source; all the acetate formed during growth remained in the medium. Although lactate fermentation continued for 12 further days and methane production from acetate and residual carbon dioxide for a further 3 weeks the thermal output of these metabolic processes was below the level of detection in the calorimeter.

For cells of *B.stearothermophilus*, growing anaerobically in glucose-limited medium (0·03%) with nitrate as the electron acceptor, the shape of the *p-t* trace depends on the temperature of growth and the concentration of nitrate (Reiling and Zuber 1983). Nitrate was not incorporated into cellular material; ammonium chloride provided the nitrogen source. At low nitrate concentrations ($\leqslant 6\,\text{mmol dm}^{-3}$) growth was predominantly nitrate-controlled; the power increased to a maximum value and decreased rapidly with the exhaustion of nitrate and nitrite and the cessation of growth (Figure 4.4). Lactate and acetate, produced during growth, attained near constant values in the stationary phase, some residual glucose was also present. At higher nitrate concentrations, growth was glucose-controlled and the total biomass was markedly greater than at low nitrate concentrations.

Under conditions of glucose excess ($\leqslant 6\,\text{mmol dm}^{-3}$ nitrate) the end oxidants nitrate and nitrite are reduced successively to gaseous nitrogenous end products of the denitrification chain ($NO_3^- \rightarrow NO_2^- \rightarrow N_2O$) with the maximum possible energy gain. At this point, after the exhaustion of the electron acceptors, the heat production decreases rapidly to zero because any residual glucose and breakdown products cannot be metabolized anaerobically. At higher nitrate concentrations ($\geqslant 8\,\text{mmol dm}^{-3}$), i.e. conditions of glucose limitation, the electron flow from glucose metabolism is transferred

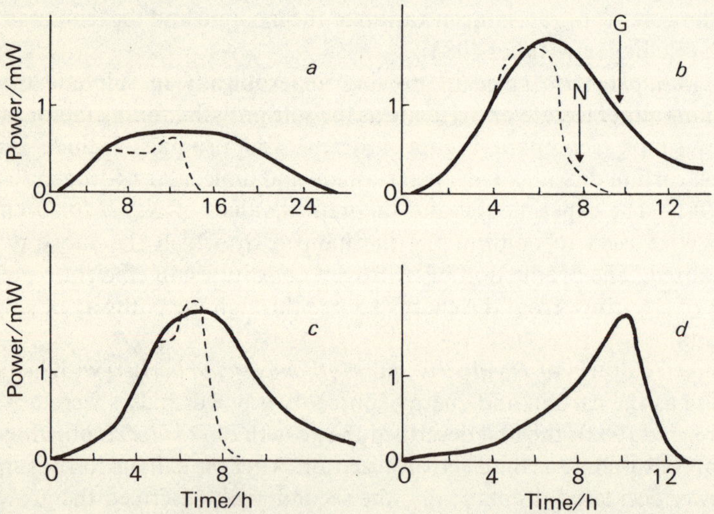

Figure 4.4. Heat production during the growth of *B.stearothermophilus* in glucose medium (0·03% w/v) at different temperatures with different concentrations of nitrate as electron acceptor (after Reiling and Zuber 1983). ----- $2\,\text{mmol dm}^{-3}$ nitrate; ——— $8\,\text{mmol dm}^{-3}$ nitrate. *a* 47 °C; *b* 57 °C; *c* 62·5 °C; *d* 57 °C with nitrite as electron acceptor.

N↓ exhaustion of nitrate; G↓ exhaustion of glucose.

completely, yielding nitrite. As a result of the presence of this residual electron acceptor further metabolic processes (e.g. acetate metabolism) can occur. Thus after glucose exhaustion there will be a prolongation of power output, resulting in the bell-shaped *p-t* trace. The heat of production of biomass during the anaerobic growth of *B.stearothermophilus* based on nitrate reduction is exceptionally high when compared to that of other organisms (Table 4.1).

The *p-t* trace of cells of *K.aerogenes* (conditioned to growth in medium of low phosphate concentration) growing under conditions of phosphate-limitation (Figure 4.5) is quite unlike those for cells growing in phosphate-sufficient medium (Bowden and James 1986b). At a concentration of 18 μmol dm^{-3} phosphate three phases of growth can be recognized from the shape of the power- and carbon dioxide-time traces: (*a*) a period of exponential growth and heat production during which free phosphate becomes exhausted, (*b*) a period of slower growth with constant power and carbon dioxide output during which glucose becomes exhausted and (*c*) a final period of minimal growth during which acetate is metabolized with low heat and carbon dioxide production. Mass balances have been established for each phase of growth and for overall growth. Overall, as a result of the stressful conditions of growth the total energy recovered was 96·6% (a decrease in the energy stored as biomass and a marked increase in waste energy as compared to growth in phosphate-sufficient medium) with only 3·4% available for biosynthetic, maintenance, storage in macromolecules and other purposes. It appears that energy is only stored under favourable conditions when surplus substrates and/or secondary metabolites are available for processing into compounds for storage; these compounds can provide energy

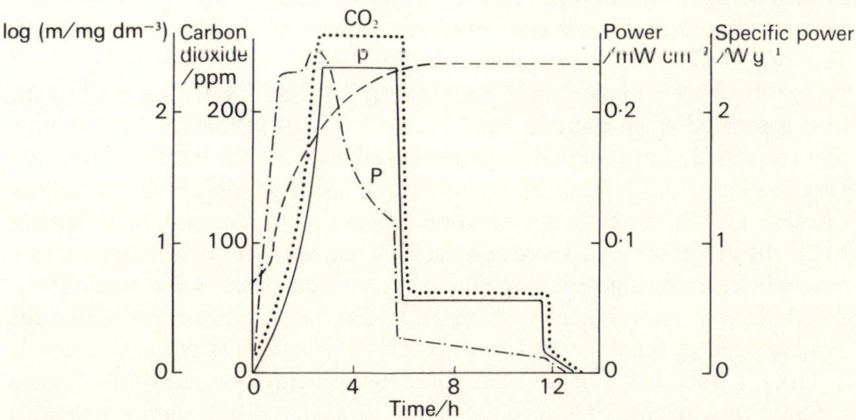

Figure 4.5. Power-time, specific power and carbon dioxide profiles for the growth of *K.aerogenes* (NCTC 418) in phosphate-limited medium (18 μmol dm^{-3}). ----- log (biomass); ——— power; —·—·— specific power; ········ carbon dioxide.

when required for growth under conditions of stress caused by nutrient deprivation or adverse environmental conditions. The storage of such compounds is less likely under conditions of phosphate limitation than under phosphate sufficiency; in support of this is the size of the ATP pool of stationary state cells: 0·75 μmol g^{-1} for phosphate-limited cells compared with 2·5 μmol g^{-1} for standard cells. Similar arguments are applied to the three individual stages of growth where it is found that the energy stored and the ATP pool are greatest in the initial phase of growth where there is no substrate limitation, becoming less first with phosphate exhaustion and then with glucose exhaustion.

The formation of biomass is an overall exothermic process, the enthalpy of formation (thermal yield) of anaerobically grown cells is greater than that for aerobic cells (Table 4.1). For normal growth, with glucose as substrate, $\Delta_p H$ is in the range $-9{\cdot}5$ to -14 kJ g^{-1} irrespective of the organism; when the cells are under stress, e.g. at higher temperatures (Nichols and James 1981a), the value is less. The enthalpy of metabolism, $\Delta_{met}H$, for anaerobic growth is very much greater than for aerobic growth, suggesting that anaerobic growth is less efficient than aerobic growth, more energy being required for bacterial synthesis; the values for a given substrate are remarkably constant for a range of organisms despite the possibility of different pathways for the metabolism of the substrate. Under stressful conditions of growth, e.g. the presence of antibacterial agents, there is a marked decrease in the value of $\Delta_{met}H$. Calculated values of $\Delta_{met}H$, which depend on the elemental analysis of cells, a measured value of Y_s and the correctness of the assumptions about the metabolic processes occurring, are generally in good agreement with those obtained experimentally.

The limiting values of $\Delta_{met}H$ for a range of carbon substrates were obtained by measuring the heat output of 5×10^9 cells of *E.coli* in very low concentrations of substrates under a controlled atmosphere of air, oxygen or nitrogen (Lovrien *et al* 1980). At the concentrations used the metabolism and heat generation were complete in 5–10 minutes. The limiting values express the heat generated when there is rapid metabolism but little growth. Although there was little, if any, actual growth and cell division, the limiting values are very similar to those reported for other organisms growing in batch culture (Table 4.1). The rate of heat generation depends on the adaptation of the cells to the substrate; for unadapted cells, at least one or two doubling times elapse before there is any appreciable heat output. *Pseudomonas putida*, used for the metabolism of naphthalene and 2-methylnaphthalene, conserves less of the available energy than does *Acinetobacter* metabolizing benzene and toluene.

The *p-t* trace for *Cellulomonas* sp 21399 growing aerobically in defined medium with glucose or amorphous cellulose as the carbon source (extended in time for 40 h) exhibits a second heat pulse probably due to the oxidation of intermediate products of fermentation (Dermoun *et al* 1984). The maximum power output of cells grown with crystalline cellulose as the substrate was 8 times less than that on amorphous cellulose. It has been established that the hydrolysis product of cellulose is consumed in the same way as that of glucose and further that the limiting factor to the biodegradability of cellulose is the breakdown of the polymeric substrate (Dermoun and Belaich 1985).

The shape of the *p-t* traces of an undefined mixed culture of soil organisms (bell shaped with secondary heat pulses extending for up to 40 h) during the degradation of a range of sugars changed with the temperature of growth. The total amount of heat generated, 14–15 kJ (g sugar)$^{-1}$, however, was constant for all the sugars in the temperature range 30–48 °C (Yamano and Takahashi 1983). Values for the apparent activation energy and Gibbs energy of activation for each of the sugars are reported. Although it is important to have information of heat changes during growth at different temperatures it must be realized that different metabolic porcesses may become rate-limiting at the different temperatures. Further, in a mixed culture the dependence of growth of one organism on the metabolism of others may also change with temperature.

There are very few reports of thermal measurements with continuous cultures (yeasts: Brettel *et al* 1972, 1981a,b,c; Carduso-Duarte *et al* 1976; bacteria: James and Djavan 1980, 1981, 1982; Ishikawa and Shoda 1983; Leiseifer and Schleser 1983), in spite of the fact that the corresponding thermodynamic data are necessary for studies of open systems and for certain problems in biotechnology. A chemostat provides the ideal conditions for a study of the metabolism of cells in a steady state; under these conditions the power output should be invariant with time so long as constant flow through the calorimeter is maintained. Problems associated with aeration, flow rate through the calorimeter and temperature control have now been eliminated and the combination of a chemostat with a flow calorimeter has been achieved and shown to give reproducible results (James and Djavan 1981; Leiseifer and Schleser 1983).

For cells of *K.aerogenes* growing in glucose-limited medium the power at constant dilution rate (D h^{-1}) is independent of the time. There is a steady increase in the power at low dilution rates ($< 0{\cdot}6$ h^{-1}) but at higher dilution rates ($D = 0{\cdot}65$–$1{\cdot}0$ h^{-1}) the power is constant ($0{\cdot}156 \pm 0{\cdot}013$ mW cm^{-3} with a biomass of 0·372 mg cm^{-3}). For a meaningful comparison of the thermal properties of cells in their steady states under different conditions of growth, values of $\Delta_g H$ (the enthalpy change associated with the formation of 1 g of cells during a generation time) are calculated. The generation time was chosen as the standard time interval, rather than a set period, e.g. 1 h (Ishikawa and Shoda 1983), because this is the time when similar changes in population etc. occur under different conditions of growth. The value of $\Delta_g H$ is independent of substrate concentration, but increases with dilution rate attaining a constant value at high dilution rates (Figure 4.6); $\Delta_g H$ increases linearly with the molar yield coefficient. In contrast for cells of *E.coli*, the power output in glucose-limited medium (11·1 mmol dm^{-3}) increases with

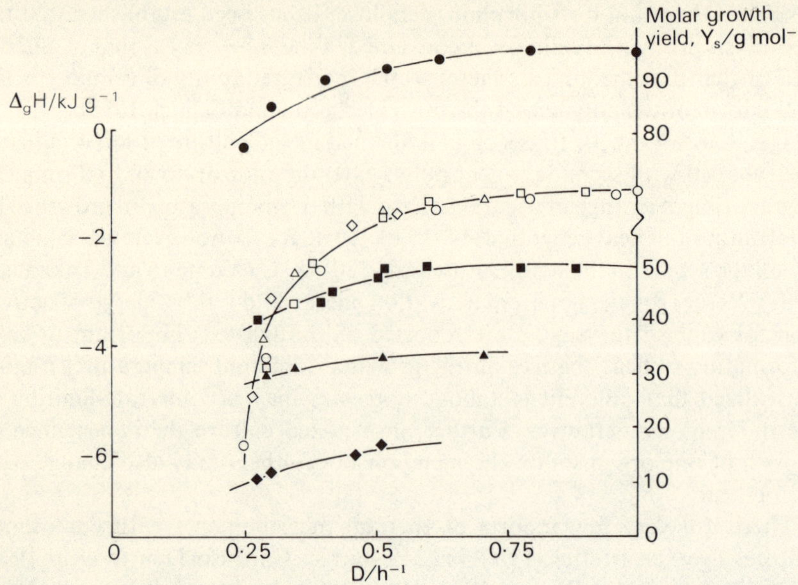

Figure 4.6. Specific heat output per generation time and molar growth yield as a function of dilution rate for *K.aerogenes* growing in defined medium with different limiting carbon sources (after Djavan and James 1980).

	$\Delta_g H$	Y_s
Glucose (4·0 mmol dm^{-3})	○	●
Glycerol (7·8 mmol dm^{-3})	□	■
Pyruvate (8·2 mmol dm^{-3})	△	▲
Acetate (12·0 mmol dm^{-3})	◇	◆

dilution rate to $D = 0{\cdot}5\,h^{-1}$, thereafter falling dramatically on account of wash-out; large amounts of acetate accumulate under these conditions. Over the limited dilution range the specific heat production rate (comparable to $\Delta_g H$) is a linear function of D (Ishikawa and Shoda 1983).

At a fixed dilution rate, $\Delta_g H$ for *K.aerogenes* is the same for growth in glucose, glycerol, pyruvate or acetate as the sole carbon and energy source; the molar growth yields are markedly different (Figure 4.6). The value of $\Delta_g H$ for cells growing in glucose-sufficient media (i.e. N- or Mg-limiting) is much lower than that in glucose-limiting media; acetate accumulated under these growth conditions. Under anaerobic conditions the molar growth yield in glucose is about one-third of the comparable value for cells growing aerobically whilst $\Delta_g H^{an}$ is twice (i.e. more negative) and the rate of evolution of carbon dioxide 2·5 times that for cells growing aerobically. In contrast to growth under aerobic conditions, acetate was produced in large amounts. It is concluded from calculations involving yield and $\Delta_g H$ data that the amount of energy lost (as heat) per mol of glucose during glycolysis is the same under aerobic and anaerobic conditions (Djavan 1980; James and Djavan 1981).

From double reciprocal plots of Y_s against D and $\Delta_g H$ against D^{-1} values for the growth rate dependent and independent maintenance coefficients in glucose and glycerol have been determined; two values for each are reported, one for values of $D < 0{\cdot}52\ h^{-1}$ the other for $D > 0{\cdot}52\ h^{-1}$ (Djavan and James 1980). The values are in good agreement with those of Nichols and James (1981b) for glucose and with those of Erickson *et al* (1979) for glycerol.

4.4.2 *Thermal Data from Cells Growing in Rich Media*

The power-time traces for organisms growing in complex media (brain heart infusion, Columbia broth, Oxoid nutrient broth, blood culture broth etc.) often with added vitamins, are more complex than those recorded during growth in defined medium with a growth-limiting substrate (Kallings and Hoffner 1977; Fujita *et al* 1978; Mardh *et al* 1978; Semenitz 1978a; Monk 1979; Herman *et al* 1980). Detailed interpretation in terms of changes in metabolic processes is difficult, if not impossible, because there is no known growth-limiting substrate, there is minimal control of environmental conditions and little or no information on associated growth parameters. In general, the power increases during exponential growth; the maximum power may be attained during the growth phase or well into the stationary phase. The *p-t* trace may indicate diauxic growth with two well separated thermal events. In rich media the total biomass produced is greater than in defined media and growth ceases as a result of the accumulation of toxic products and/or 'over-crowding' of the culture. After growth ceases considerable amounts of heat are evolved as a result of the metabolism by resting cells of residual unidentified metabolites or secondary intermediates produced during growth. Considerable amounts of carbon dioxide are produced during this time (Bunker 1985) but, unlike growth in C-limited medium, the carbon dioxide-time trace does not parallel the *p t* trace. Tables of maximum power generated and total heat evolved have no significance in the absence of growth yields and other associated data.

The shape and position of the *p-t* trace is probably more dependent on the history of the parent inoculum (age, method of storage, viability), the stirring/aeration conditions and the composition of the growth medium (Figures 4.7, 4.8) rather than on the species; the richer the medium the more complex the trace (Schaarschmidt and Lamprecht 1976; Bunker 1985). Changes in the *p-t* trace caused by a small change in the growth medium may not be a simple response to a different preferred energy substrate but to the limitation of growth factors which the cells may be required to synthesize during some stage of growth. Regulation of metabolism of one compound by varying concentrations of another may also be involved. It is therefore apparent that, despite numerous claims to the contrary (Russell *et al* 1975; Schaarschmidt and Lamprecht 1976), the microcalorimetric method has severe limitations for

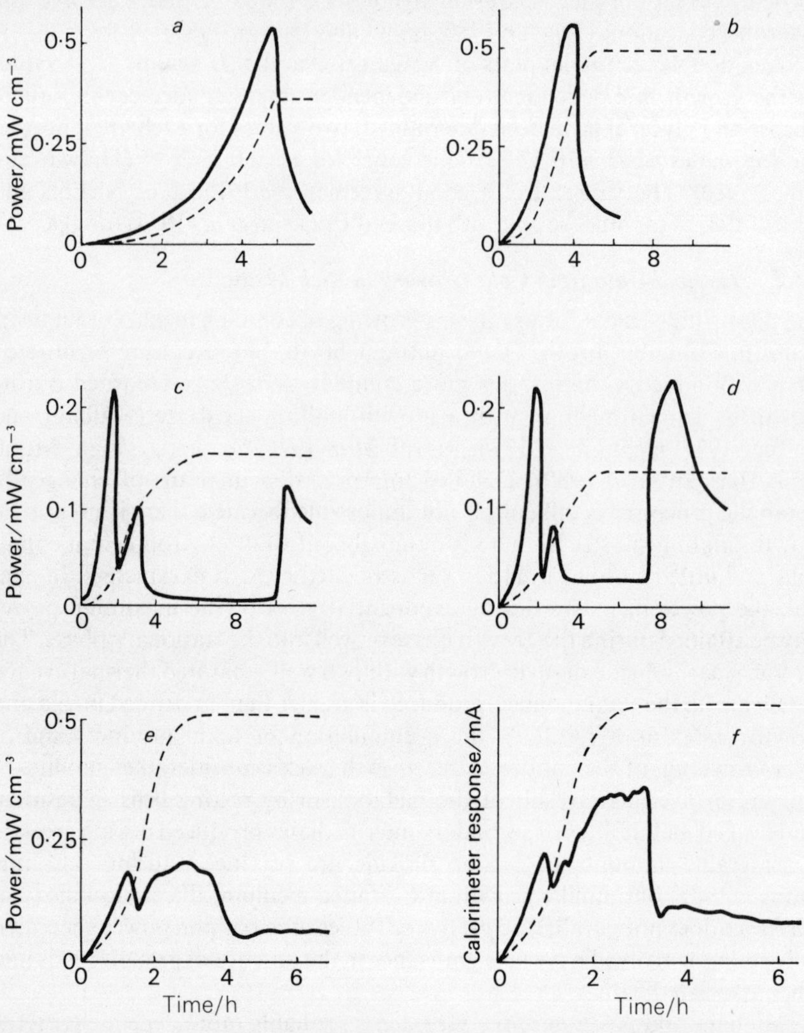

Figure 4.7. Power-time traces for the growth of *K.aerogenes* in different media, flow calorimetry (- - - - - growth curve). NCTC 418 growing in: *a* glucose-limited medium; *b* glucose-limited medium containing 5% w/v Oxoid nutrient medium; *c* glucose-limited medium containing 10% w/v Oxoid nutrient medium; *d* Oxoid nutrient medium (after Bunker 1985). *e* 6004 growing in Columbia broth (after Semenitz and Casey 1983). *f* 225 growing in Columbia broth (after Semenitz *et al* 1982).

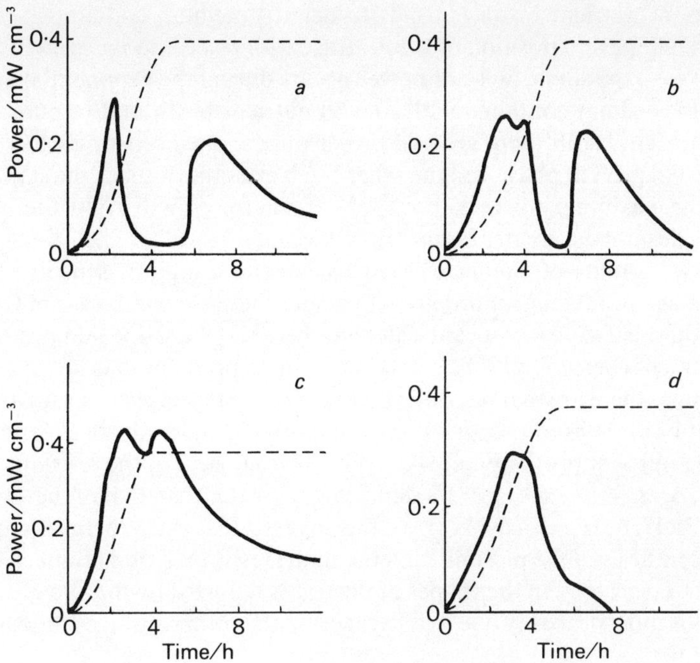

Figure 4.8. Power-time traces for the growth of *S.aureus* in nutrient broth under different aeration and stirring conditions; flow calorimetry (after Bunker 1985). (- - - - - growth curve). 13137 (antibiotic resistant): *a* Low stirrer rate; *c* high stirrer rate. Ps.47 (antibiotic sensitive): *b* low stirrer rate; *d* high stirrer rate (compare Figure 4.2 *d*).

bacterial identification; it is extremely inflexible and makes rigid demands on all operations (Newell 1980).

For strains of *E.coli* growing aerobically (flow calorimeter) the general shape of the *p-t* traces is broadly similar for different strains, but there are variations in the fine detail (Semenitz *et al* 1982, 1983). There is an initial peak of heat production followed successively by a partial trough, when the energy for anabolic processes exceeds that produced by the catabolic processes, a second peak and a descending flank with a shoulder. There is a prolonged period of relatively low heat production without biomass formation. There is no significant difference between the *p-t* traces for ampicillin-sensitive and -resistant strains. The maximum power of cells growing in a batch cell and the total heat produced are considerably less than for cells growing in a fermentor (Mardh *et al* 1976a), suggesting that anaerobic growth is less efficient than aerobic growth.

For three strains of *K.aerogenes* growing aerobically in different complex media (Figure 4.7) the *p-t* traces are strikingly different. The addition of small amounts of nutrient broth (Oxoid) to defined medium containing glucose alters the shape and position of the *p-t* trace with respect to the growth curve. At 5% (w/v) the single pulse of power occurs during the exponential growth phase. In medium containing 10% (w/v) nutrient broth and in pure (standard) nutrient broth there are two power pulses, one occurring early in the exponential growth phase and the other 5–6 h after the onset of the stationary phase; the maximum power is about 30% of that for growth in glucose-limited medium despite the greater biomass produced.

For two strains of *Staphylococcus aureus* (an antibiotic-sensitive and a β-lactamase-producing strain) growing under the same conditions in Columbia broth, there is no significant difference between the shape and position of the *p-t* traces (Semenitz 1978c); data on biomass production is not available. In contrast, there are marked differences in the *p-t* traces for a sensitive and a fully antibiotic resistant strain of *S.aureus* growing under identical conditions in Oxoid nutrient broth (Figure 4.8). The thermal yield of the resistant strain (13 137, $\Delta_p H = -3{\cdot}8\ \mathrm{kJ\,g^{-1}}$) is significantly greater than that of the sensitive strain (Ps.47, $\Delta_p H = -6{\cdot}5\ \mathrm{kJ\,g^{-1}}$); this suggests that extra energy is required for the synthesis of the plasmid material in the resistant strain (Bunker 1985). The difference between the shapes of the traces reported by the two groups of workers is most probably due to differences in the composition of the complex media.

Three variants of a tetracycline-resistant strain of *S.aureus* obtained by curing (a sensitive and two resistant strains, one R-plasmid-mediated and the other as a result of mutation), when grown in nutrient broth had coincident *p-t* traces (Lebek *et al* 1982).

During the growth of the luminescent bacterium *Beneckea harveyi* in a sea-water medium the initial heat flux is accompanied by very little growth; there is no light flux during this period (McIlvaine and Langerman 1977). At the point where luciferase synthesis starts, there is a marked decrease in the power. This occurs at the same time as the first doubling period of the population and it is postulated that this arrest in the power output is due to endergonic anabolic metabolism. In 'conditioned' medium, all the characteristics of the *p-t* trace are altered, the decrease in heat flux is now concurrent with the second doubling period. The power and light outputs of *Photobacterium leiognathi* increase exponentially during exponential growth and attain maximum values at the onset of the stationary phase; thereafter both decrease dramatically.

The rapid decrease in power 3 h after the commencement of growth of *Serratia marinoruba* in three different media was due to oxygen deficiency in the ampoule cell. In medium lacking an electron acceptor further growth was limited and the secondary heat output was low (Bermudex *et al* 1984).

Power-time traces, without growth curves, are reported for *Streptococcus thermophilus* and *Lactobacillus bulgaricus* (Monk 1979); *Citrobacter freundii*, *Levinea malonatica*, *L.amalonatica*, *K.pneumoniae*, *Enterobacter aerogenes*,

Hafnia alvei, Serratia marcescens, S.liquefaciens and *S.fonticola* (Herman *et al* 1980); *Enterobacter cloacae, Haemophilus influenzae, Pseudomonas aeruginosa, Streptococcus pneumoniae, Listeria monocytogenes, Salmonella typhimurium, S.enteritidis, Neisseria meningitidis* (Kallings and Hoffner 1977).

4.4.3 Thermal Data from Cells Growing in the Presence of Antibacterial Agents

The presence of an antibacterial agent in a growing culture normally results in changes in cellular metabolic processes which, in turn, may give rise to changes in the thermal properties of the culture. Thermal studies, carried out under carefully controlled environmental conditions, provide information on the interference of metabolism of growing cells, often at concentrations below the MIC; for detailed interpretation of the experimental results, data on biomass and other growth parameters must be collected at the same time. The majority of reports (reviewed extensively by Chowdhry *et al* 1983) are of a qualitative nature only, merely being descriptions of changes in the shape and position of *p-t* traces on the addition of antibiotics to cultures in the early exponential phase of growth; few also contain growth curves and observations on changes in cellular morphology and lysis of the cells. There is no common pattern to the variation of the power on the addition of an inhibitor; the change depends on the type of inhibitor and, for antibiotics, on the sensitivity or resistance of the organisms.

Metabolic inhibitors (sodium azide, 2,4-dinitrophenol, DNP, and carboxycyanid-*p*-trifluoromethyl-phenylhydrazone, FCCP) all had a pronounced effect on the *p-t* profile of cells of *K.aerogenes* growing in glucose-limited medium; that of azide was very different from those of DNP or FCCP (Nichols 1980). Heat output in the presence of the inhibitors was increased and the biomass decreased (Table 4.2).

The addition of dicyclohexyl-carbodiimide, DCCD (100 μmol) to a glucose-limited culture of *K.aerogenes* in its steady state ($D = 0{\cdot}5\,h^{-1}$) caused an increase in the rate of respiration and carbon dioxide production; this was accompanied by a 2–3 fold increase in the power. There was a 30% decrease in the biomass, indicating wasteful utilization of the carbon source. The steady state values resumed after several hours, only when the added inhibitor had been diluted out of the fermentor (Djavan 1980).

Exposure of non-β-lactamase-producing *E.coli* and *S.aureus* to benzyl penicillin and ampicillin gave rise to a 'paradoxical zone phenomenon' in which the inhibitory effect on bacterial metabolism, judged from the heat generated, was smaller when the concentration was 5–10 × MIC than at 2 × MIC and at 100–10 000 × MIC (Mardh *et al* 1979). Except at very high concentrations there was a second increase in the power. There was only a small interruption in the *p-t* trace of β-lactamase producing organisms exposed to benzyl penicillin.

When either amoxycillin, AMX (40 $\mu g\,cm^{-3}$) or clavulanic acid, CA

Table 4.2 Thermal yield, experimental and calculated values of the enthalpy of metabolism of glucose for cells of *K.aerogenes* growing aerobically in glucose-limited medium after the addition of different inhibitors during the early exponential growth phase (flow-calorimetry at 37 °C)

Inhibitor	*Final inhibitor concentration* /μmol dm^{-3}	Y_{gluc} /g g^{-1}	*Thermal yield* $-\Delta_p H$/kJ g^{-1}	$-\Delta_{met}H$/kJ mol^{-1} *Experimental*	$-\Delta_{met}H$/kJ mol^{-1} *Calculated*	*Reference*
Control	0	0·43	14·4	1038		Nichols (1980)
DNP	1000	0·33	28·7	1590		
	2000	0·25	24·6	1026		
NaN_3	8000	0·22	36·3	1426		
FCCP	10	0·43	16·1	1243		
	30	0·43	15·9	1220		
	60	0·33	19·0	1130		
	130	0·24	26·4	1140		
Antimycin A	70	0·45	13·5	1170		
Control	0	0·50	11·4	1120	1200	Bowden and James (1985c)
Nalidixic acid	60	0·45	13·4	1140	1350	
	80	0·42	14·7	1220	1450	
	180	0·41	14·6	1390	1470	
	415	0·39	18·0	1350	1540	
	690	0·29	15·7	1680	1860	
	1035	0·32	20·7	1760	1780	
	1420	0·35	18·7	2190	1670	

(10 $\mu g\,cm^{-3}$) were added separately to AMP-resistant *E.coli* cultures (in the exponential growth phase), the *p-t* and growth curves were the same as for the control (Figure 4.9). When AMX (20 $\mu g\,cm^{-3}$) and CA (10 $\mu g\,cm^{-3}$), i.e. augmentin, were added together to the growing resistant culture, the *p-t* trace showed a marked reduction in heat production and the growth curve showed bacterial lysis (Semenitz and Casey 1982; Semenitz *et al* 1983). CA had no effect on the *p-t* trace of AMP-sensitive *E.coli*, but AMX caused a marked decrease in heat production and cell lysis; the extent of both depended on the concentration of AMX. Exposure of the sensitive strain to a combination of AMX and CA (at concentrations which on their own had no effect) produced a considerable inhibition of heat production and bacterial lysis. The morphological changes undergone by AMP-resistant strains in the presence of AMX and CA were qualitatively similar to those produced in the sensitive strain by AMX alone. The effect of the clavulanic acid is due to its inhibition of β-lactamase. CA alone caused a reduction in the heat production and the turbidity of a β-lactamase-producing strain of *S.aureus*; the inhibitory effect was even more pronounced when AMX was also present; i.e. a synergistic effect.

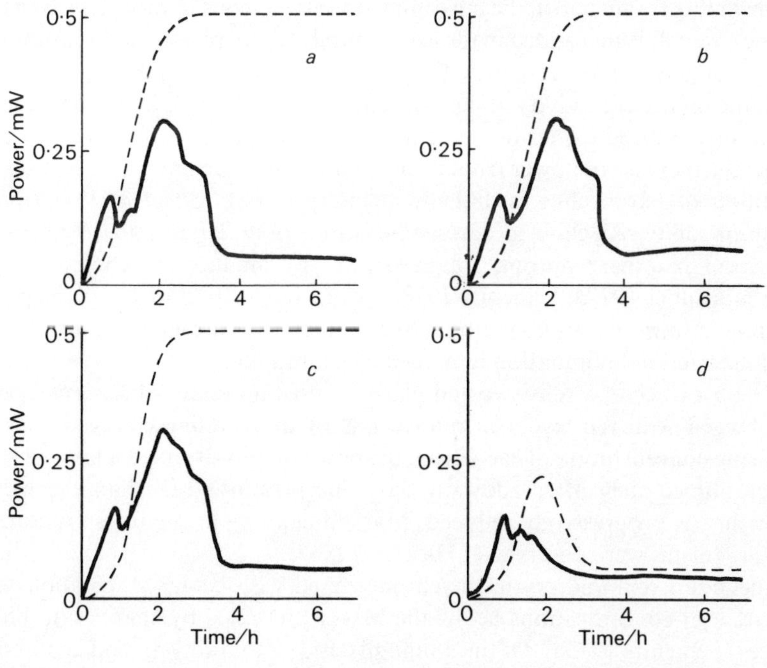

Figure 4.9. Power-time traces for the growth of *E.coli* JT39 (β-lactamase producing strain) in Columbia broth in the presence of antibacterial agents (after Semenitz and Casey 1983) (----- growth curve). *a* control; *b* clavulanic acid 10 $\mu g\,cm^{-3}$; *c* amoxycillin 40 $\mu g\,cm^{-3}$; *d* amoxycillin 20 $\mu g\,cm^{-3}$ + clavulanic acid 10 $\mu g\,cm^{-3}$.

Qualitative changes in the *p-t* traces of *E.coli* and *S.aureus* brought about on the addition of other β-lactam antibiotics are reported: ampicillin, epicillin, oxacillin and carbenicillin (Semenitz 1978c; 1980) and thienamycin (Semenitz *et al* 1981).

The changes in the *p-t* traces brought about by cefoxitin on β-lactamase producing *S.aureus* are similar to those reported for the penicillins. Cefoxitin develops maximum blocking effect on metabolism with lysis at very low concentrations (Semenitz 1979a). The *p-t* traces of *E.coli* cultured in a non-aerated growth vessel in the presence of cephalexin or cephaloridine at a concentration of 0·5 × MIC did not differ in shape from the control, but were displaced along the time axis. At a concentration of 1 × MIC there were no heat effects (Arhammer *et al* 1978). On the addition of either of these cephalosporins to *E.coli* or *S.aureus* growing in trypticase soy broth (flow calorimeter) there was a short period of increased power followed immediately by a decrease to a low constant value. The secondary increase in the power, after a time which depended on the challenging concentration, was due to the selection of resistant mutants.

At a concentration of 0·5 × MIC, four tetracyclines prolonged the lag phase of cells of *E.coli* growing in a batch calorimeter cell and delayed the onset of the heat flux for up to 8 h, the maximum power was considerably less than that of the control. When added in the exponential growth phase, all the antibiotics caused an immediate decrease in the power, but the extent and duration varied with the biological activity of the drug (Mardh *et al* 1976a). Doxycycline has a more pronounced bacteriostatic effect over a longer time than has tetracycline; the increase in power from its near constant value immediately after the addition of tetracycline is difficult to interpret (Mardh *et al* 1976b). The addition of tetracycline to a resistant strain of *S.aureus* caused an initial cessation in power output, followed later by an increase (Semenitz and Tiefenbrunner 1977a; Semenitz 1978c); while for a strain of *E.coli* the power decreased immediately to the base-line (Semenitz and Tiefenbrunner 1977b) and the bacterial population remained constant.

The *p-t* trace of a sensitive and plasmid-resistant strain of *S.aureus*, when challenged with tetracycline, minocycline or doxycycline changed with increasing concentration of each drug; the power finally attained a low constant value immediately after addition, (at concentrations less than the MIC); growth was progressively reduced. Marked changes in the ultrastructure of the organisms were observed (Lebek *et al* 1982).

Oleandomycin and erythromycin interfered with bacterial metabolism of *S.aureus* at concentrations below the MIC as revealed by changes in the *p-t* traces (Semenitz 1978a). Of the aminoglycosides, tobramycin and gentamicin influence the *p-t* traces of *E.coli* in a similar way, the power decreasing to the base-line with lysis at higher concentrations. Spectinomycin, however, produced different shaped traces; reduction in the growth rate and growth at higher concentrations gave rise to near constant power production (Semenitz 1978c, 1979b).

Cinoxacin caused a decrease in and delay in the heat effects of a culture of *E.coli* growing in trypticase soy broth; the second increase in the power is attributed to the selection of cinoxacin-resistant mutants (the MIC increased from 1·7 to 400 $\mu g\,cm^{-3}$ after 17 h exposure) (Mardh *et al* 1978).

Diphenhydramine (a local anaesthetic and antagonist of both acetylcholine and histamine) inhibits the growth of *E.coli* and *S.aureus*. The addition of diphenhydramine to growing cultures (final concentration below the MIC) interferes with cell metabolism; growth ceases and the power decreases dramatically to zero (Semenitz 1978b). *p*-Cresol has a marked effect on the *p-t* trace of *E.coli* growing in nutrient broth in a batch calorimeter; increasing concentrations cause a delay in the production of heat and a marked decrease in the total heat released (Hartung 1982).

The addition of nalidixic acid (NA) to exponentially growing cells of *K.aerogenes* in simple salts/glucose medium has a marked effect on the thermal and other growth parameters of the cells (Bowden and James 1985c). The shape of the *p-t* trace changes in a manner which is dissimilar to those reported earlier (Figure 4.10). There is no abrupt decrease in the power on the addition of NA and since the cells are not killed or lysed the *p-t* trace eventually reattains the base line. The time taken for the power to become zero increases with increasing concentration due to the formation of acetate which is broken down in the stationary phase with the liberation of much heat. With increasing NA concentration there is a decrease in the yield and in the energy stored in biomass, but the energy wasted as heat increases and in consequence less energy is stored in macro-molecular species during the growth of cells under stress. The thermal yield of the cells increases with increased stress, indicating a decrease in the efficiency of the metabolic processes (Table 4.2).

The variation of specific power (at the time of exhaustion of glucose) with concentration shows two distinct regions of antibacterial potency. At low concentrations ($<80\,\mu mol\,dm^{-3}$) the potency per mol is constant; at higher concentrations the potency is again constant but about 18 times less than at the lower concentrations. Using conventional microbiological techniques a similar dual-dose related potency for NA against Gram-negative organisms has been reported (Crumplin and Smith 1975). After repeated growth of the cells in the presence of low concentrations of NA, the yield and waste heat of the 'trained' cells were the same irrespective of whether growth occurred in the presence or absence of the antibiotic. This suggests that during training, the original stress imposed on the metabolic processes has been overcome and they are now capable of operating with similar efficiency in the presence or absence of the drug.

Some platinum group metal complexes (PGMC) are antibacterial at low concentrations against a range of organisms growing in different nutrient media (Dwyer *et al* 1969). In a microcalorimetric study of the action of a number of PGMC on the metabolism of *K.aerogenes* or *S.aureus*, a freshly prepared solution of the complex was added to an aerobically growing culture in a fermentor (final concentration $10^{-4}\,mol\,dm^{-3}$) during the early part of

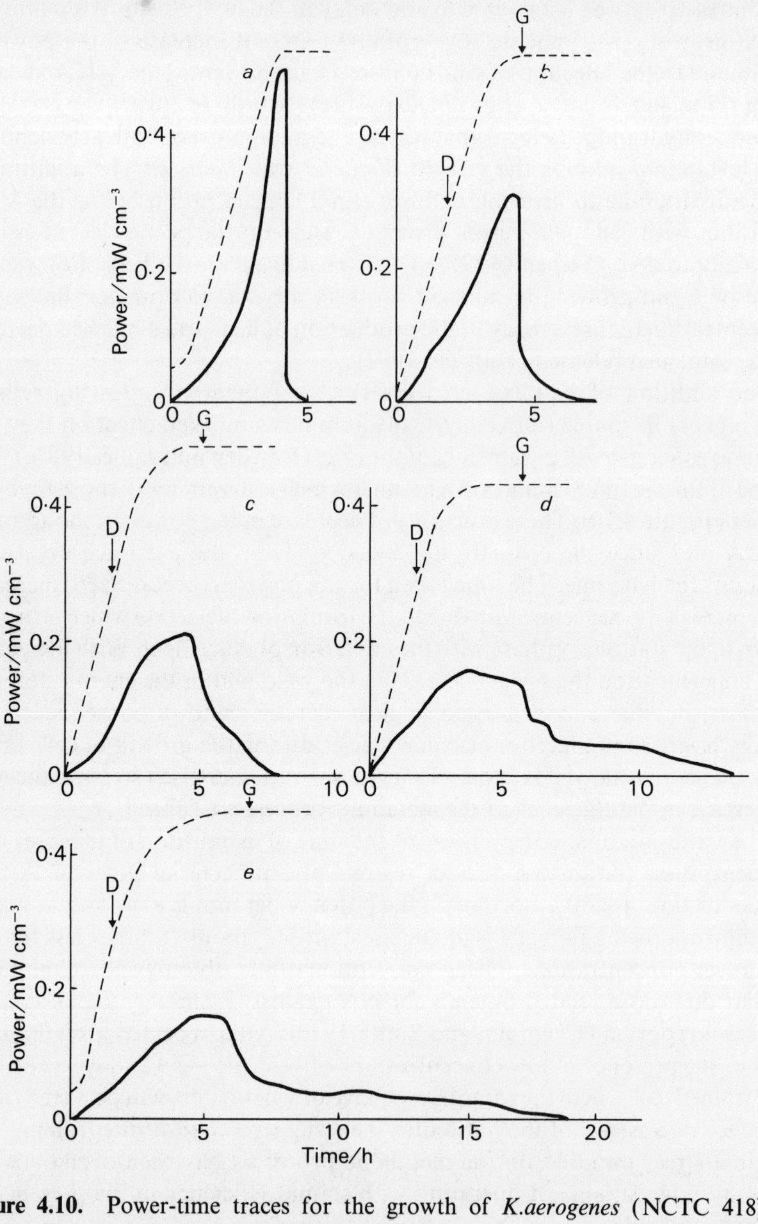

Figure 4.10. Power-time traces for the growth of *K.aerogenes* (NCTC 418), in glucose-limited medium when nalidixic acid is added during exponential growth (after Bowden and James 1985c)

(----- growth curve). D↓ introduction of NA into culture; G↓ exhaustion of glucose. Final concentration of NA in culture: *a* 0; *b* 60; *c* 180; *d* 690 and *e* 1420 μmol dm^{-3}.

exponential growth. The power, biomass and carbon dioxide outputs were monitored. For a large number of the complexes, although there is no significant change in the shape or position of the *p-t* trace, small differences in the thermal yield indicated detectable changes in the metabolism. The addition of complex PGMC 1 to *K.aerogenes* growing in glucose-limited defined medium caused an immediate cessation of growth and decrease of power to zero (Figure 4.11). There was negligible growth and power if PGMC 1 was added to the medium prior to inoculation. When nutrient broth was included in the glucose-limited medium (final concentration 5% w/v), the response on the addition of PGMC 1 was markedly different (Figure 4.11). At a similar concentration the complex did not have any pronounced effect on the *p-t* trace or growth of *S.aureus* in nutrient broth; however at higher concentrations there was an immediate reduction in the power (Figure 4.11). It was concluded that interaction between PGMC 1 and some components present in nutrient broth was responsible for the apparent reduction in the activity of the complex (Bunker 1985).

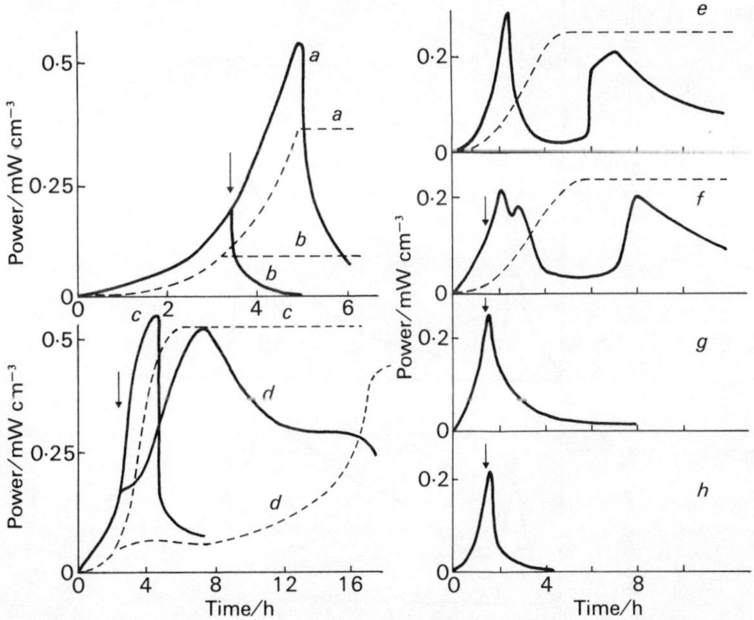

Figure 4.11. Power-time traces for the growth of *K.aerogenes* (NCTC 418) and *S.aureus* (13137) when PGMC 1 is added during exponential growth (after Bunker 1985) (- - - - growth curve). ↓ introduction of complex. *K.aerogenes* growing in glucose-limited medium: *a* control; *b* after addition of complex (final concn. 10^{-4} mol dm^{-3}). *K.aerogenes* growing in glucose-limited medium supplemented with Oxoid nutrient broth (5% w/v): *c* control; *d* after addition of complex (10^{-4} mol dm^{-3}). *S.aureus* growing in Oxoid nutrient broth: *e* control and after the addition of PGMC 1 at various final concentrations; *f* 10^{-4}; *g* 10^{-3}; *h* 10^{-2} mol dm^{-3}.

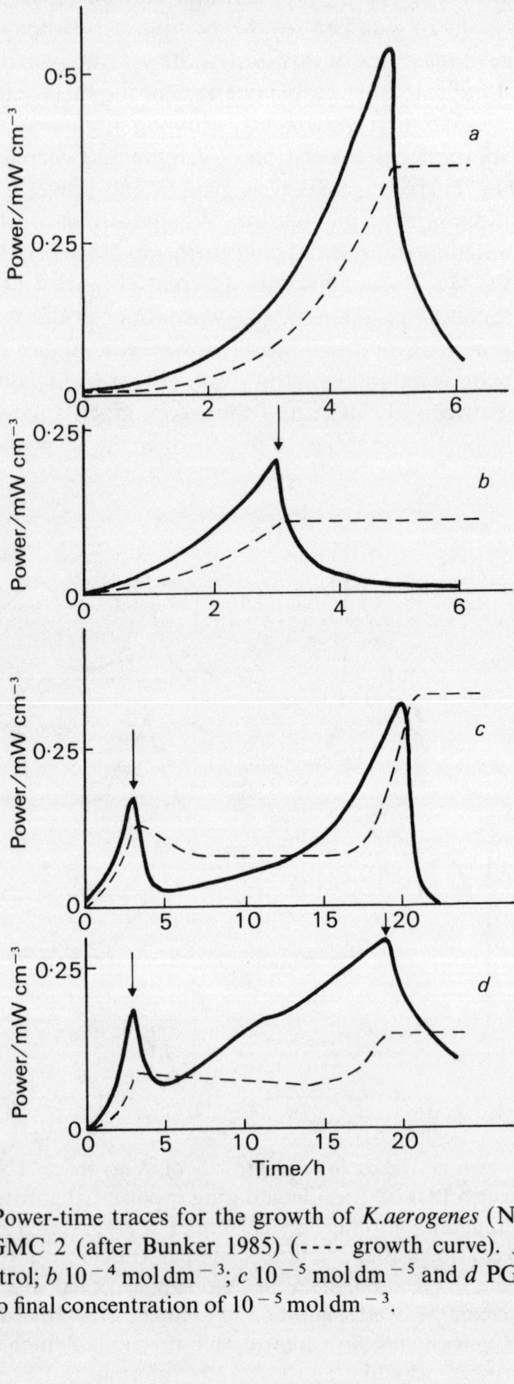

Figure 4.12. Power-time traces for the growth of *K.aerogenes* (NCTC 418) on the addition of PGMC 2 (after Bunker 1985) (---- growth curve). ↓ introduction of complex. *a* control; *b* 10^{-4} mol dm^{-3}; *c* 10^{-5} mol dm^{-5} and *d* PGMC 2 added at 3 and 19 h each to final concentration of 10^{-5} mol dm^{-3}.

When added to cultures of *K.aerogenes* growing in glucose-limited medium, freshly prepared solutions of three complexes, PGMC 2, 3 and 4, all had similar effects on the growth at final concentrations of 10^{-4} and 10^{-5} mol dm^{-3} (Figure 4.12). At the higher concentration all three complexes caused immediate cessation of power and biomass production. At 10^{-5} mol dm^{-3}, the complexes caused an immediate reduction in power and arrest of growth but after several hours power and biomass production resumed. Spectroscopic evidence indicates that there were changes in the nature of the complexes in solution which resulted in a reduction of the activity, thereby allowing growth to resume. A further addition of the complex (to the same final concentration) after the resumption of growth had a similar effect on the power and biomass as after the first addition; thus resumption of growth is not due to the growth of resistant bacteria. Confirmation of change in the active species of PGMC was obtained by adding a solution of this compound which has been kept at 37 °C for 6 h before injection into the culture; similar changes in the *p-t* trace and growth were observed but now regrowth occurred after a shorter interval (Bunker 1985).

4.5 Applications of Thermal Studies

On account of its non-specific nature microcalorimetry has certain advantages over conventional microbiological techniques in the study of the metabolism of pure and mixed cultures. Changes in metabolism, as a result of changed environmental conditions revealed by thermal changes can be detected earlier and at lower cell populations than by conventional techniques. Studies of the degradation of heterogeneous solid substrates, such as straw, and the microbial activity of soils and sewage caused by the presence of mixed cultures can be undertaken and the extent of action monitored.

The degradation of cellulose during the fermentation of straw, an abundant and cheap substrate, to produce such industrial products as alcohols, acids and gases, is often limited because of the presence of lignin and hemicellulose. The biodegradability of straw can be increased by mechanical, biological, chemical or radiation pretreatment. Anaerobic microcalorimetry in conjunction with gas chromatography has been used to follow the degradation of straw by an undefined mixed bacterial culture (obtained from a pile of weeds or from a rice field in the Camargue) and to estimate the efficiency of the pretreatment (Fardeau *et al* 1980). The *p-t* traces recorded over a 7-day period of cellulose degradation of straw in a simple buffered salts medium are reproducible. The quantity of heat produced and the carbon recovered during this time are proportional to the initial amount of straw in the calorimeter cell.

The limiting factor of straw degradation is the cellulytic activity; the addition of cellulase to the growth medium increases the amount of straw degraded by a factor of 4 and reduces the time of fermentation (Fardeau *et al*

1981). The enthalpy of metabolism associated with the fermentation of each cellulose monomer ($C_6H_{10}O_5$) was 120 kJ, a value close to that for the metabolism of hexoses by pure cultures (cf. Table 4.1). Pretreatment of the straw with potassium hydroxide (Plasse *et al* 1980; Partos *et al* 1983) and sterilization by γ-irradiation (Partos *et al* 1982) improved the straw degradation from 6 to 40%. After 7 days' fermentation only traces of methane were detectable in the fermentation products; after this time methane production occurs for about 1 month. The thermogenic phase recorded corresponds to the acidogenic phase which is not limited by the overall cellulase activity of the culture.

The microbial metabolic activity of sewage digesting systems, monitored by the heat production of samples from an active anaerobic digestor, is dependent on the concentration of organic compounds present and is correlated with the dehydrogenase activity. An advantage of monitoring by microcalorimetry is that an assessment can be made as to whether heavy metal ions acting as inhibitors of metabolism are reversible, as for the presence of Zn^{++} or irreversible as for Fe^{+++} (Redl and Tiefenbrunner 1981).

The 'biological half-time' of slimicides, which indicates the breakdown of anti-slime agents with respect to their biological activity, have been measured with a flow-microcalorimeter. Using this technique it has proved possible to differentiate between bactericidal and bacteriostatic effects (a most important consideration from both the economic and ecological points of view) and to monitor the development of resistant strains of microflora in paper-making water as a result of prolonged dosage with slimicides (Harju-Jeanty 1982).

In soils there is a continual interchange between small active bacterial populations which increase, peak and then fade as different energy sources, growth factors etc. successively appear and become exhausted. Although many methods have been used in an attempt to quantitatively characterize the activity of the microflora, no single method has proved sufficiently informative, accurate and technically convenient (Stotzky 1972; Sparling 1981). The low specificity of the calorimetry method is an asset in such studies. In an ampoule calorimeter the heat evolved from a soil sample originates mainly from the metabolic activity of viable cells, from biochemical systems of dead organisms in the soil matrix with a very small amount ($<$ 10%) from processes unrelated to biological systems (Mortensen *et al* 1973).

The addition of nutrients (e.g. glucose, ammonium sulphate) stimulates microbial activity and results in an immediate increase in heat production; this is followed by a decrease in the heat flux on the depletion of oxygen or an accumulation of toxic products (e.g. carbon dioxide). Sterilization of the sample by autoclaving results in an up to 4-fold decrease in the heat production (Ljungholm *et al* 1979a). Acid treatment of the soil has little effect on the microbial activity, but when supplied with an extra energy source the heat effects are significantly lower than those of the untreated controls (Ljungholm *et al* 1979b; 1980). Although there are still many problems to be solved, the calorimetric technique can be used to monitor changes in soil caused by the addition of fertilizers, biocides, etc.

Although there are many reports of changed thermal properties brought about by the addition of antibiotics to growing bacterial cultures, there are few reports of quantitative assays based on microcalorimetry for antibacterial agents. In contrast, flow microcalorimetry, using *Saccharomyces cerevisiae*, provides a rapid, sensitive and reproducible assay for polyene antibiotics (Beezer and Chowdhry 1980). The interaction of *Streptococcus faecalis*, growing in a complex medium, with penicillin G or tetracycline produces changes in the *p-t* trace which, it is claimed, can be used to assay low levels of the antibiotics in serum (Jensen *et al* 1976).

Changes in the thermal properties of a strain of *E.coli* growing in a semi-defined medium, brought about by the addition of a range of penicillins at concentrations in excess of 1 × MIC, have enabled the construction of dose-response curves (Beezer *et al* 1980). Various response parameters were tested including (*a*) the time for the *p-t* trace to reattain the base line level, (*b*) the relative peak size in the presence and absence of the antibiotic and (*c*) the time taken from the addition of the antibiotic to achieve maximum power (<35 min). In developing a suitable microcalorimetric bioassay, the change in the *p-t* trace on the addition of the antibacterial agents must be reproducible and quantitative and the medium standardized (preferably simple and defined) to avoid interaction of the antibacterial agent with components of the growth medium.

Values of the phenol coefficients of a range of biocides, measured by flow-microcalorimetry, compare favourably with those obtained by conventional techniques (Perry *et al* 1980; Beezer *et al* 1981, 1983). The constant power output of respiring cells of *E.coli* (10^8 organisms cm^{-3}) in buffered glucose medium is reduced on the addition of the biocide; the phenol coefficient can be obtained by comparing the height of the peak power output in the presence of the biocide with that of a phenol control. The time of assay is reduced from several days to less than 1 h.

References

Arhammer M, Mardh P-A, Ripa T and Andersson K-E (1978) *Acta Pathol Microbiol Scand* **86B**, 59–65

Bayne-Jones S and Rhees H S (1929) *J Bacteriol* **17**, 123–40

Beezer A E (1977) In: Lamprecht I, Schaarschmidt B (eds) *Applications of Calorimetry in the Life Sciences*. Berlin, de Gruyter, pp 109–18

Beezer A E (ed) (1980) *Biologial Microcalorimetry*, London, Academic Press

Beezer A E and Chowdhry B Z (1980) In: Beezer A E (ed) *Biological Microcalorimetry*. London, Academic Press, pp 195–246

Beezer A E, Miles R J, Shaw E J and Vickerstaff L (1980) *Experientia* **36**, 1051–2

Beezer A E, Hunter W H and Storey D E (1981) *J Pharm Pharmacol* **33**, 65–8

Beezer A E, Hunter W H and Storey D E (1983) *J Pharm Pharmacol* **35**, 406–7

Belaich A and Belaich J-P (1976) *J Bacteriol* **125**, 19–24

Belaich J-P (1963) *Comp Rend Soc Biol* **157**, 317–25

Belaich J-P (1980) In: Beezer A E (ed) *Biological Microcalorimetry*. London, Academic Press, pp 1–42

Belaich J-P, Senez J C and Murgier M (1968) *J Bacteriol* **95**, 1750–7

Bermudez J, Loren J G, Vinas M and Wagensberg J (1984) *Microbios Lett* **27**, 61–5

Bovinet P (1964) *Perfectionnements aux methodes d'etalonnage et d'emploi du Microcalorimetre E.CALVET. Utilisation en microbiologie*. PhD Thesis, L'Universite d'Aix-Marseille

Bowden C P P (1983) *Variations in the Energy Output and ATP Pool During the Growth of Klebsiella aerogenes in Defined Media*. PhD Thesis, London University

Bowden C P P and James A M (1985a) *Microbios* **43**, 93–105

Bowden C P P and James A M (1985b) *Microbios* **44**, 75–85

Bowden C P P and James A M (1985*c*) *Microbios* **44**, 201–16

Brettel R, Corti L, Lamprecht I and Schaarschmidt B (1972) *Studia Biophysica* **34**, 71–6

Brettel R, Lamprecht I and Schaarschmidt B (1981a) *Thermochim Acta* **49**, 53–61.

Brettel R, Lamprecht I and Schaarschmidt B (1981b) *Eur J Appl Microbiol Biotechnol* **11**, 205–11

Brettel R, Lamprecht I and Schaarschmidt B (1981c) *Eur J Appl Microbiol Biotechnol* **11**, 212–5

Bunker J C (1985) *The Effects of Some Platinum Group Metal Complexes on Bacterial Growth*. PhD Thesis, London University

Cardoso-Duarte J M, Marinho M J and van Uden N (1976) In: Dean A C R *et al* (eds) *Continuous Culture* 6. *Applications and New Fields*. Chichester, E. Horwood, pp 40–8

Chowdhry B Z, Beezer A E and Greenhow E J (1983) *Talanta* **30**, 208–43

Cooney C L (1968) *Measurement of Heat Evolution During Fermentation*. MSc Thesis, MIT

Cooney C L, Wang D I C and Mateles R I (1968) *Biotechnol Bioeng* **11**, 269–81

Crumplin C C and Smith J T (1975) *Antimicrob Agents Chemother* **5**, 251–61

Dermoun Z and Belaich J-P (1979) *J Bacteriol* **140**, 377–80

Dermoun Z and Belaich J-P (1980) *J Bacteriol* **143**, 742–6

Dermoun Z and Belaich J-P (1985) *Biotechnol Bioeng* **27**,1005–11

Dermoun Z, Boussand R, Cotten D and Belaich J-P (1985) *Biotechnol Bioeng* **27**, 996–1004

Dermoun Z, Gaudin C and Belaich J-P (1984) *Appl Microbiol Technol* **19**, 281–7

Djavan A (1980) *Microcalorimetric Studies of Chemostat Cultures of Klebsiella aerogenes*. PhD Thesis, London University

Djavan A and James A M (1980) *Biotechnol Lett* **2**, 303–8

Dwyer F, Reid K, Schulman A, Laycock G M and Dixson S (1969) *Aust J Exp Biol Med* **42**, 203–18

Erickson L E, Minkevich I G and Eroshin V K (1979) *Biotech Bioeng* **21**, 575–80

Eriksson R and Wadsö I (1971) In: Broda E *et al* (eds) *Proc 1st Eur Biophys Congr*. Wein, Med Akad, pp 319–27

Eriksson R and Holme T (1973) *Biotech Bioeng Symp* **4**, 581–90

Fardeau M L, Plasse F and Belaich J-P (1980) *Eur J Appl Microbiol Biotechnol* **10**, 133–43

Fardeau M L, Plasse F, Partos J and Belaich J-P (1981) In: Palz W, Chartier P (eds) *Energy from Biomass. 1st E C Conference*. London, Applied Sciences, pp 380–5
Forrest W W (1972) In: Norris J R, Ribbons D W (eds) *Methods in Microbiology*. New York, Academic Press, pp 285–317
Fujita T, Monk P R and Wadsö I (1978) *J Dairy Res* **45**, 457–63
Glanser M, Ban S and Kniewald J (1979) *Process Biochem* **14**, 17–25
Gordon A S and Millero F J (1980) *Thalassia Jugosl* **16**, 405–24
Gordon A S, Millero F J and Gerchakov S M (1982) *Appl Environ Microbiol* **44**, 402–9
Grangetto A (1963) PhD Thesis, L'Universite d'Aix, Marseille
Gustafsson L (1979) *Arch Microbiol* **120**, 15–23
Gustafsson L and Lindman B (1977) *FEMS Microbiol Lett* **1**, 227–30
Harju-Jeanty P (1982) *Appita* **36**, 26–31
Hartung J (1982) *Fortschr Veterinaermed* **35**, 329-33
Heinritz B, Stoll P and Glombitz A (1983) *Acta Biotechnol* **3**, 83–7
Herman J P, Jakubczak E, Izard D and LeClerc H (1980) *Can J Microbiol* **26**, 413–9
Herman J P, Jakubczak B, Belaich J-P and LeClerc H (1982) *Eur J Appl Microbiol Biotechnol* **14**, 212–5
Ishikawa Y and Shoda M (1983) *Biotechnol Bioeng* **25**, 1817–28
Ishikawa Y, Shoda M and Maruyama H (1981a) *Biotechnol Bioeng* **23**, 2629–40
Ishikawa Y, Nonoyama Y and Shoda M (1981b) *Biotechnol Bioeng* **23**, 2825–36
James A M and Djavan A (1980) *Microbios* **29**, 171–83
James A M and Djavan A (1981) *Microbios* **30**, 163–70
James A M and Djavan A (1982) *Microbios* **34**, 17–20
Jensen T E, Hansen L D, Eatough D J, Sagers R D, Izatt R M and Christensen J J (1976) *Thermochim Acta* **17**, 65–71
Kallings L O and Hoffner S (1977) LKB Application Note No. 309, 1–6
Lamprecht I and Schaarschmidt B (1973) *Bull Soc Chim France* 1200–1
Lamprecht I and Schaarschmidt B (eds) (1977) *Application of Calorimetry in Life Sciences*. Berlin, de Gruyter
Lebek G, Semenitz E, Gstraunthaler G and Pfaller W (1982) *Immun Infekt* **10**, 193–201
Leiseifer H P and Schleser G H (1983) *Zeit Naturforsch* [*C*] **38**, 259–67
Ljungholm K, Noren B, Skold R and Wadsö I (1979a) *Oikos* **33**, 15–23
Ljungholm K, Noren B and Wadsö I (1979b) *Oikos* **33**, 24–30
Ljungholm K, Noren B and Odham G (1980) *Oikos* **34**, 98–102
Lock M A and Ford T E (1983) *Appl Environ Microbiol* **46**, 463–7
Lock M A and Ford T E (1985) *Appl Environ Microbiol* **49**, 408–12
Lovrien R, Jorgensen G, Ma M K and Sund W E (1980) *Biotech Bioeng* **22**, 1249–70
Luong J H T and Volesky B (1980) *Can J Chem Eng* **58**, 497–504
Luong J H T and Volesky B (1982a) *Can J Chem Eng* **60**, 163–7
Luong J H T and Volesky B (1982b) *Eur J Appl Microbiol Biotech* **16**, 28–34
Mardh P-A, Ripa T, Andersson K-E and Wadsö I (1976a) *Antimicrob Agents Chemother* **10**, 604–9
Mardh P-A, Andersson K-E, Ripa T and Wadsö I (1976b) *Scand J Infect Dis Suppl* **9**, 12–16
Mardh P-A, Arhammer M, Coleen S and Andersson K-E (1978) *J Antimicrob Chemother* **4**, 73–8
Mardh P-A, Arhammer M and Andersson K-E (1979) *Chemotherapy* **25**, 106–16
McIlvaine P and Langerman N (1977) *Biophys J* **17**, 17–25

Minkevich I G and Eroshin V K (1973) *Folia Microbiol* **18**, 376–80
Monk P R (1979) *J Dairy Res* **46**, 485–96
Mortensen U, Noren B and Wadsö I (1973) *Bull Ecol Res Commun* **17**, 189–97
Mou D G and Cooney C L (1976) *Biotechnol Bioeng* **18**, 1371–92
Newell R D (1980) In: Beezer A E (ed) *Biological Calorimetry*. London, Academic Press, pp 163–86
Nichols S C (1980) *Microcalorimetric Study of Energy Changes During Bacterial Growth*. PhD Thesis, London University
Nichols S C and James A M (1980) *Microbios* **29**, 95–104
Nichols S C and James A M (1981a) *Microbios* **31**, 153–60
Nichols S C and James A M (1981b) *Biotechnol Lett* **3**, 119–24
Nichols S C and James A M (1983) *Microbios* **38**, 51–63
Nichols S C, Prichard F E and James A M (1979) *Microbios* **25**, 187–203
Partos J, Fardeau M-L and Belaich J-P (1982) In: Strub A *et al* (eds) *Energy from Biomass, 2nd E.C. Conf.* London, Applied Sciences, pp 1006–11
Partos J, Fardeau M-L and Belaich J-P (1983) *Eur J Appl Microbiol Biotechnol* **18**, 308–14
Payne W J (1970) *Ann Rev Microbiol* **24**, 17–52
Perry B F, Beezer A E and Miles R J (1980) *Microbios* **29**, 81–7
Plasse F, Fardeau M-L and Belaich J-P (1980) *Biotechnol Lett* **2**, 11–16
Poole R K and Haddock B A (1975) *FEBS Lett* **58**, 249–53
Redl B and Tiefenbrunner F (1981) *Eur J Appl Microbiol Biotechnol* **12**, 234–8
Reiling H E and Zuber H (1983) *Arch Microbiol* **136**, 243–53
Schaarschmidt B and Lamprecht I (1973) *Experientia* **29**, 505–6
Schaarschmidt B and Lamprecht I (1976) *Experientia* **32**, 1230–4
Schaarschmidt B, Zotin A I, Brettel R and Lamprecht I (1975) *Arch Microbiol* **105**, 13–16
Semenitz E (1978a) *J Antimicrob Chemother* **4**, 455–7
Semenitz E (1978b) *Wein Klin Wochenschr* **90**, 710–14
Semenitz E (1978c) *Immun Infekt* **6**, 260–6
Semenitz E (1979a) *Zeit Klin Ther Infekt* **7**, 68–70
Semenitz E (1979b) *Zentralbl Bakteriol* [*Orig A*] **243**, 372–80
Semenitz E (1980) *Thermochim Acta* **40**, 99–107
Semenitz E and Tiefenbrunner F (1977a) In: Lamprecht I, Schaarschmidt B (eds) *Application of Calorimetry in Life Sciences*. Berlin, de Gruyter, pp 251–60
Semenitz E and Tiefenbrunner F (1977b) *Drug Res* **27**, 2247–51
Semenitz E and Casey P A (1982) In: Croydon E A P, Michel M F (eds) *Proc Eur Symp Scheveningen, The Netherlands*. Excerpta Medica 1983, Amsterdam, Elsevier, pp 68–76
Semenitz E, Gstraunthaler G and Pfaller W (1981) *Current Chemotherapy and Immunotherapy. Proc 12th Int Cong of Chemotherapy, Florence*. Washington DC, The American Society for Microbiology, pp 708–10
Semenitz E, Gstraunthaler G and Pfaller W (1982) In: Dettli L *et al* (eds) *Cephalosporin, Hahnenklee Symp No 9.* Basel, Rocke, pp 47–63
Semenitz E, Casey P A, Pfaller W and Gstraunthaler G (1983) *Chemotherapy* **29**, 192–207
Sparling G P (1981) *Soil Biol Biochem* **13**, 93–8
Stotzky G (1972) *Crit Rev Microbiol* **2**, 59–137

Swartz J R (1985) *Int Biotechnol Lab* **3**, 36–43

Traore A S, Fardeau M-L, Hatchikian C E, LeGall J and Belaich J-P (1983) *Appl Environ Microbiol* **46**, 1152–6

Traore A S, Hatchikian C E, Belaich J-P and LeGall J (1981) *J Bacteriol* **145**, 191–9

Wadsö I (1970) *Q Rev Biophys* **3**, 383–427

Wadsö I (1975) In: Pain R H, Smith B J (eds) *New Techniques in Biophysics and Cell Biology*. London, Wiley, pp 85–126

Wadsö I (1980) In: Beezer A E (ed) *Biological Microcalorimetry*. London, Academic Press, pp 247–74

Wadsö I (1983) *Pure Appl Chem* **55**, 515–28

Yamano H and Takahashi K (1983) *Agric Biol Chem* **43**, 1493–9

Zotin A I and Lamprecht I (eds) (1976) *Thermodynamics of Biological Processes*. Russian Edition, Moscow, Nauka; English Edition (1978) Berlin, de Gruyter

Chapter 5

Growth and Metabolism of Yeasts

R. J. Miles, A. E. Beezer and B. F. Perry

5.1 Introduction

Yeasts are fungi which are predominantly unicellular and which reproduce vegetatively (i.e. asexually) by budding and/or fission (transverse division). They are a diverse group and include members of the order Endomycetales (e.g. *Saccharomyces* and *Schizosaccharomyces* spp.) in which sexual reproduction leading to the formation of ascospores may take place, ballistospore-producing yeasts, and yeasts producing neither ballistospores nor ascospores and for which no sexual phase has been demonstrated. Yeasts are generally maintained as pure strains of vegetatively grown cells. Under suitable cultural conditions some yeasts (e.g. *Candida* sp.) may form a branched mycelium.

Yeasts are widely distributed and may be isolated from soil, animal excreta and from plants, particularly in nectar and on the surface of fruits where high concentrations of sugars are found. Some species are pathogenic to man and/or animals; *Cryptococcus neoformans* is a causative agent of pulmonary disease and meningitis, and *Candida albicans* may be associated with infections of the mucous membranes, skin and lungs.

Economically, yeasts are the most important group of microorganisms exploited by man. Their major commercial uses are in baking, as human and animal food supplements and in ethanol production, both in alcoholic beverages and for fuel and synthetic purposes. The principal genera of commercial interest are *Saccharomyces* and *Candida*. Yeasts are also of commercial importance as food-spoilage organisms, particularly where the pH value is below 5·5 (Walker and Ayres 1970).

In addition to commercial applications, *Saccharomyces cerevisiae* (*Sacch. cerevisiae*) has been widely used to study the genetics, growth and biochemistry of eukaryotic cells; it has been suggested (Rose and Harrison 1971) that this organism should become a focus of attention as regards eukaryotic cell studies, in order to obtain a complete understanding of its organization.

The nutritional requirements of yeast have been reviewed by Suomalainen and Oura (1971). The sources of carbon and energy used vary according to the species of yeast and are of importance in yeast classification and identification (Kreger-van Rij 1969). Anaerobically, *Sacch. cerevisiae* is able to utilize glucose, fructose, mannose, galactose, sucrose and maltose; aerobically, it may additionally utilize such relatively simple carbon sources as, e.g. acetic, oxaloacetic and α-ketoglutaric acids. Substrates used by other yeasts include polysaccharides, polyols and hydrocarbons.

Sugars are metabolized by either the Embden-Meyerhof or hexose monophosphate pathways. Sugar transport has been reviewed by Eddy (1982). In *Sacch. cerevisiae*, there is substantial evidence that the transport of monosaccharides involves a carrier in the membrane, which acts so as to equilibrate sugar concentration on either side of the membrane, i.e. the transport system is one of facilitated diffusion. However, proton symports, implying an active transport, have been demonstrated for maltose, sucrose and α-methylglucoside uptake by *Sacch. fragilis*. Glucose uptake was not associated with a proton symport in this yeast, although monosaccharide uptake by a respiration-dependent proton symport occurs in *Rhodotorula gracilis*. In some yeasts disaccharides may be hydrolysed by extracellular enzymes prior to uptake, e.g. sucrose by *Sacch. cerevisiae* and maltose by a strain of *Schizosaccharomyces pombe* (*Schizosacch. pombe*).

The growth-factor requirements of yeasts are diverse. The most common requirements are for biotin, pantothenic acid, inositol, thiamine, nicotinic acid and pyridoxine. Additionally, under anaerobic conditions, yeasts require a source of sterol; it has been suggested that a major function of molecular oxygen in yeast growth is for sterol synthesis (David 1974).

Nutrients other than the carbon/energy source and growth factors may be provided as inorganic salts, as in the defined media of Wickerham (1951) and Brandli (1980). All yeasts are able to utilize ammonium ions as a source of nitrogen, and some may utilize nitrate. Growth rate is high in media with mixtures of amino acids, although growth rate and cell yield may be low, or growth may not occur, where a single amino acid is the sole source of nitrogen.

The effect of physical parameters (temperature, pH and water activity) on yeast growth and metabolism has been reviewed by Dawes (1976). The minimum growth temperature for most yeast strains is close to 0 °C, although obligately pyschrophilic strains are able to grow at temperatures as low as −7 °C. With the exception of these strains, the maximum temperature for growth is between 30 and 47 °C; however, growth rate, cell-yield and product formation are not always highest at the maximum growth temperature. Most yeasts grow best at acid pH; the pH optimum for *Sacch. cerevisiae* is about 4·5.

The majority of yeast strains are capable of aerobic respiration; however, oxygen utilization may be dependent upon sugar (glucose) concentration. In glucose-sensitive yeast strains, glucose represses respiration by inhibiting synthesis of respiratory enzymes. Thus, under aerobic conditions, *Sacch. cerevisiae* and other glucose-sensitive yeasts may initially metabolize glucose to ethanol; when glucose is exhausted, ethanol is oxidized to carbon dioxide. In contrast, many yeast strains, including some strains belonging to *Candida* and *Rhodotorula*, do not produce ethanol and are insensitive to free glucose. These strains cannot be grown in the absence of free oxygen. The control of glucose metabolism in yeast has been reviewed by Fiechter *et al* (1981). In anaerobically grown yeast (i.e. glucose-sensitive strains) the synthesis of respiratory enzymes is reduced and the structure of mitochondria is altered (Matile *et al* 1969). Cell yield for the energy substance is lower, reflecting the reduced ATP yield of fermentation compared to that of respiration.

There have been a substantial number of microcalorimetric studies in which yeasts have been used. In this review an attempt will be made to assess the impact of such studies on our understanding of yeast growth/metabolism, and upon the application of microcalorimetry to strain identification/characterization, evaluation of complex media and to microbiological assays. Studies of yeast growth/metabolism will be discussed in relation to the microbiological culture techniques employed.

5.2 Microcalorimetric Studies of Yeast Growth/Metabolism

5.2.1 Batch-culture Studies

A batch culture consists of an homogeneous suspension of organisms in an initial limited amount of nutrient medium. After the lag period growth occurs at the maximum rate and finally ceases through lack of nutrient, accumulation of an inhibitory product(s) or some change in the physical environment. After the biomass reaches its maximum, it may remain at a 'stationary' level before ultimately declining as a result of maintenance metabolism or autolysis (Pirt 1975).

There have been many microcalorimetric studies of the growth of *Sacch. cerevisiae* in batch culture. 'Batch' calorimeter vessels and flow microcalorimetric systems have been used. These should ideally give the same results; however, the control of 'batch' systems may present practical difficulties (Wadsö 1980). Earlier studies have been described by Lamprecht (1980) and accordingly in this review emphasis will be given to more fundamental and recent publications. In glucose-limited media under anaerobic conditions with low inoculum levels, power-time (*p-t*) curves show an initial lag followed by an exponential increase in power and a relatively rapid decline due to glucose exhaustion. Total heat evolved closely follows the curve for biomass production (Schaarschmidt *et al* 1975). Under aerobic conditions *p-t* curves may be biphasic; the first phase corresponding to the fermentation of glucose to ethanol and the second, in which the heat evolved is much greater, to the oxidation of ethanol to carbon dioxide. Schaarschmidt and Lamprecht (1977) were able to correlate changes in power output with disappearance of glucose and the formation and subsequent utilization of ethanol. In the first phase power increased exponentially, representing exponential growth of the culture. In the second phase the manner of increase in power was complex, presumably reflecting the complexity of the process of adaptation to respiration. Heat evolved during the first phase of growth may, however, be greater in the presence of oxygen than in its absence, suggesting some oxygen utilization during the first phase (Brettel 1977a). Under aerobic conditions, triphasic *p-t* curves have also been reported by Brettel *et al* (1980a); power in the final phase was postulated to be due to oxidation of acetate, formed during the initial phase.

In conducting aerobic microcalorimetric studies of yeast growth it is essential that the oxygen solution rate of the culture system is sufficient to meet oxygen demand which may be considerable (Pirt 1975). Newell (1975) has contrasted the vastly different *p-t* curves obtained for *Sacch. fragilis* under oxygenated and inadequately aerated culture, and Brettel (1977a) has similarly shown substantial differences in *p-t* curves of *Sacch. cerevisiae* grown aerobically in a stirred-batch calorimeter cell and in an intensively agitated and aerated fermentor monitored by a flow microcalorimeter. The literature includes data which may have been erroneously interpreted due to inadequate aeration of cultures (Newell 1980). Increasing dissolved oxygen concentration by increasing the pressure of oxygen above cultures may lead to severe inhibition of growth and heat evolution (Lamprecht and Schaarschmidt 1973; Schaarschmidt *et al* 1974). Toxic effects of high oxygen concentrations are well known, even amongst aerobic and facultatively anaerobic organisms (Gottlieb 1971; Morris 1975).

The effect of inoculum density and glucose concentration on the anaerobic growth of commercial strains of *Sacch. cerevisiae* in a defined minimal medium has been investigated in a flow-microcalorimetric system by Perry *et al* (1979, 1983a,b). To ensure comparability between experiments inocula of each yeast strain used were derived from a single growth of cells stored in liquid nitrogen. With inoculum densities up to 7×10^5 colony forming units (cfu cm^{-3}, *p-t* curves showed a period of exponential increase in power. The time taken to reach maximum power output (p_{max}), at a fixed glucose concentration, increased with decreasing inoculum density. For each strain, plots of this peak-time versus log (inoculum density) were linear, reflecting the exponential nature of cell growth, each halving of inoculum density increasing peak-time by one generation time. Thus, lag time was not affected by inoculum density in the range studied. When inoculum density was increased to $3{\cdot}5 \times 10^6$ cfu cm^{-3}, *p-t* curves lost their typical shape and power increased linearly with time. These findings are in accord with those of Lamprecht (1969) and Hashimoto and Takahashi (1982).

The value of p_{max} increased with increasing glucose concentration. In glucose-limited medium p_{max} occurred immediately prior to glucose exhaustion and was proportional to cell concentration at that time. Cell concentration would comprise the initial population plus progeny cells. Since the population of progeny cells will be approximately proportional to initial glucose concentration, plots of p_{max} versus initial glucose concentration would be expected to be linear, with an intercept on the power axis representing the power output attributable to the initial population of cells. This relationship was observed in the experiments of Perry *et al* (1983b); however, Lamprecht *et al* (1971), and Schaarschmidt and Lamprecht (1977) observed p_{max} to be a quadratic function of glucose concentration. At a high initial glucose concentration (100 mmol dm^{-3}) Perry *et al* (1983a) observed that *p-t* curves showed periodic fluctuations in power output. Similar fluctuations with *Klebsiella aerogenes* (Nichols *et al* 1979) have been interpreted as being due to utilization of secondary metabolites. However, Perry *et al* (1983a) attributed their results to a physical disturbance created by the formation of carbon dioxide bubbles, which they were able to reproduce artificially by introducing gas bubbles into the culture flow.

In the experiments of Perry *et al* (1983a) power declined sharply after the exhaustion of glucose, but there was an anomalous and unexplained residual power. In addition, values for the enthalpy change for metabolism ($\Delta_{met}H$; Belaich *et al* 1981) of glucose (at the point of glucose exhaustion) were high compared to the theoretical value of 135 kJ mol^{-1} for the conversion of glucose to ethanol plus carbon dioxide. Thus, there may be significant enthalpy changes associated with the utilization of medium components other than glucose, the transfer of excreted cell products to the medium etc. Lamprecht (1980) quotes experimental values of $\Delta_{met}H$ for glucose of 64–130 kJ mol^{-1} under anaerobic conditions; however, with one exception (Battley 1960) these determinations were made in complex media. In aerobic batch culture, in a complex medium, a value for $\Delta_{met}H$ for glucose of 2000 kJ mol^{-1} has been reported (Battley 1960), which is to be compared with a theoretical value of 2872 kJ mol^{-1} for the conversion of glucose to carbon dioxide plus water. Loureiro-Dias and Arrabaca (1982), found a close correlation between theoretical and experimental enthalpy changes for the growth of a respiration-deficient strain of *Sacch. cerevisiae* under anaerobic conditions. Under aerobic conditions, however, the enthalpy change was higher; the difference was shown to be due to cyanide-resistant respiration. The determination of enthalpies of metabolism might be best determined in chemostat continuous-culture (5.2.3). In the steady-state, power output, biomass, residual substrate and product concentrations, the level of reserve carbohydrate material in cells etc. can be determined in more fully defined conditions than is possible in batch culture where all of these parameters vary with time.

There have been a number of microcalorimetric investigations in which sugar combinations or sugars other than glucose have been used. The aim has been either to differentiate between strains of yeast (5.3.1), to obtain kinetic data (5.2.2) or to study patterns of utilization of sugars. Under anaerobic conditions, *p-t* curves of a wild yeast strain were diauxic, i.e. showed two distinct peaks of power in media with glucose plus galactose or sucrose or maltose, but only a single peak with glucose plus mannose or fructose (Schaarschmidt and Lamprecht 1977, 1978). The presence of two peaks reflects sequential utilization of the sugars. It appeared that glucose inhibited the synthesis of enzymes or transport components of the maltose pathway, since only a single peak of power was observed when a mutant yeast strain, said to be constitutive for the maltose-utilizing pathway, was grown on glucose plus maltose medium. However, under similar conditions, Perry *et al* (1983b) obtained monophasic *p-t* curves for the growth of various commercial strains of *Sacch. cerevisiae* on glucose plus maltose medium. Thus, the diversity of yeast strains is emphasized.

The utilization of ethanol and acetate by *Sacch. cerevisiae* in aerobic batch culture has also been studied calorimetrically. Brettel *et al* (1981a) derived a value of $\Delta_{met}H$ of 902 kJ mol^{-1} for ethanol utilization and established an energy balance within experimental precision. As expected, power increased exponentially with time, and was monophasic. Battley (1960) reported a $\Delta_{met}H$ value for acetate of 678 kJ mol^{-1}.

While the majority of yeast studies in batch culture have used media limited by the concentration of the carbon/energy source, Lamprecht and Meggers (1969) have followed power output by *Sacch. cerevisiae* in a medium in which glucose was in excess and the concentration of the nitrogen source limited growth. Power rose to a plateau value, and then remained fairly constant for several hours before declining, indicating that although growth stopped upon exhaustion of the nitrogen source (at the start of the plateau) metabolism of glucose continued independently of growth and at a rate comparable to that by growing cells.

The effect of sodium chloride concentration on the growth of the halotolerant yeast *Debaromyces hansenii* in batch culture has also been investigated by flow microcalorimetry (Gustafsson and Norkrans 1976; Gustafsson 1979). Cultures were grown aerobically in a glucose-limited medium; increasing salinity reduced growth yield, p_{max} and the time to reach p_{max}. The total heat evolved, also dependent on salinity, was 1690 and 1200 kJ mol^{-1} at 2·7 mol dm^{-3} and 4 mmol dm^{-3} sodium chloride, respectively. The specific rate of heat production (i.e. the rate per unit biomass) decreased during growth; this decrease was relatively small in the presence of 4 mmol dm^{-3} and 0·68 mol dm^{-3} sodium chloride, but was more than tenfold with 2·7 mol dm^{-3} sodium chloride. It was suggested that the high initial specific power output was associated with the development of a structural organization for salt tolerance which was more difficult to obtain at high salinities.

5.2.2 Kinetics of Sugar Utilization Determined by Analysis of Batch Culture Power-Time Curves

Belaich *et al* (1968) analysed *p-t* curves of anaerobic glucose-limited growth of *Zymomonas mobilis* and of a 'petite' (respiration-deficient) mutant of *Sacch. cerevisiae*. They assumed that at any time power output was proportional to the rate of glucose consumption and that the enthalpy change was proportional to the biomass produced and to the glucose utilized. Thus, integration of the area under *p-t* curves allowed determination of biomass and residual glucose concentration (S), and the power output per unit biomass was proportional to the specific rate of glucose utilization (V). Plots of $1/V$ versus $1/S$ were essentially linear, indicating that Michaelis kinetics were observed and allowing determination of K_m and V_{max} values for glucose utilization. In later publications, K_m and V_{max} values were similarly determined for fructose, galactose and maltose utilization by the 'petite' *Sacch. cerevisiae* strain (Murgier and Belaich 1971) and for glucose, fructose, mannose, galactose, sucrose and maltose utilization by a wild yeast strain (Schaarschmidt and Lamprecht 1978). In this latter study growth on various pairs of sugars was also observed. For the sugar combinations sucrose plus maltose and sucrose plus glucose diauxic *p-t* curves were obtained, and plots of $1/V$ versus $1/S$ gave curves with two linear portions, each representing the utilization of a single sugar.

Values of K_m and V_{max} derived in microcalorimetric studies (Table 5.1) will be determined by the rate-limiting step in the metabolism of the limiting substrate. The values derived are dependent upon the substrate used, suggesting that it is transport or some other early metabolic step, unique to the utilization of each substrate which is limiting. Belaich *et al* (1968) argued that the derived values of K_m and V_{max} related to transport rather than to subsequent metabolism. Thus, deviations from linearity in plots of $1/V$ versus $1/S$ at high S for *Sacch. cerevisiae* with galactose (Murgier and Belaich 1971) and for *Zymomonas mobilis* with glucose (Belaich *et al* 1968) may indicate that at a high sugar concentration, in these cases, transport rate does not limit the overall rate of metabolism.

The kinetic analysis of Belaich *et al* (1968) has been extended by Itoh and Takahashi (1984). They determined values of the maximum specific growth rate (μ_{max}) and K_m for the growth of a baking strain of *Sacch. cerevisiae* in glucose-limited media at temperatures in the range 16–40 °C. The value of μ_{max} varied substantially with temperature; the plot of log (μ_{max}) versus $1/T$ was linear for temperatures between 16 and 30 °C, suggesting the possibility that cell growth is kinetically limited by a single biochemical process in this temperature range and allowing calculation of an activation energy for growth of $56{\cdot}3 \pm 5{\cdot}1$ kJ mol^{-1}.

The optimum growth temperature was around 33 °C. Interestingly, the enthalpy change per unit glucose reached a minimum at this temperature and, it was suggested, could be regarded as an index to express the energy efficiency in biosynthesis; the lower the efficiency, the higher the enthalpy change during

Table 5.1 Kinetic data for sugar utilization by various strains of *Sacch. cerevisiae*. Data derived by analysis of batch culture *p-t* curves as described in the text

Yeast strain	*Sugar*	K_m /mmol dm^{-3}	V_{max} /µmol (g dry wt of cells)$^{-1}$ min^{-1}	*Note*
respiratory-deficient strain				
log-phase cells	glucose	3·2	600	a
retardation-phase cells	glucose	1·8	600	a
	fructose	4·0	550	b
	galactose	3·2	550	b
wild-strain				
	glucose	1·8	240	c
	fructose	4·2	280	c
	mannose	3·4	180	c
	galactose	1·2	140	c
	sucrose	1·7	120	c
	maltose	4·0	110	c
bakery-strain				
	glucose	1·1–1·6		d

Notes: (a) anaerobic conditions, Belaich *et al* (1968); (b) anaerobic conditions, Murgier and Belaich (1971); (c) anaerobic conditions, Schaarschmidt and Lamprecht (1978); (d) aerobic conditions, K_m varied with temperature (16–40 °C), Itoh and Takahashi (1984).

growth. Values of K_m for glucose utilization were, however, almost unchanged with temperature (Table 5.1). Assuming that the derived values of K_m relate to transport of the limiting substrate, this indicates that glucose uptake by *Sacch. cerevisiae* during growth is accompanied by only a small enthalpy change ($-6{\cdot}2 \pm 2{\cdot}5$ kJ mol^{-1}); free-energy change and entropy change were determined as $-16{\cdot}4 \pm 0{\cdot}2$ kJ mol^{-1} and $34{\cdot}2 \pm 8{\cdot}4$ J K^{-1} mol^{-1}, respectively.

5.2.3 Chemostat Continuous-culture Studies

Chemostat cultures are homogeneous suspensions of organisms into which medium is fed at a constant rate and the culture is harvested at the same rate so that the culture volume remains constant. Provided that the dilution rate (D), the medium flow rate divided by the volume of the chemostat culture, is less than the maximum specific growth rate (μ_{max}), then the chemostat culture will come to an equilibium, the 'steady state', where all measurable parameters of the culture are constant and the specific growth rate (μ) is equal to D. Culturing microorganisms in a chemostat enables greater control of their growth and behaviour than is possible in a batch culture. In particular, the growth rate may be varied with little change in the environment other than the concentration of the growth-limiting substrate (in nutrient-limited media) or growth rate may be fixed at a constant value while the environment is altered, thus allowing effects due to growth rate and to environmental factors to be separated. The ability to control growth rate is of fundamental importance, as growth rate affects many aspects of cell metabolism, structure and composition. Chemostat cultures also offer advantages in productivity compared with batch cultures.

Flow microcalorimetry has in recent years been employed in a number of chemostat growth studies. These studies have generally used defined nutrient-limited media. In such media, μ varies with the concentration of limiting substrate (S) according to the empirically derived equation:

$$\mu = \frac{\mu_{\text{max}} S}{K_{\text{m}} + S}$$

where K_{m} is the saturation constant. Values of K_{m} are nutrient-dependent and vary widely with the strain of microorganism; Pirt (1975) quotes K_{m} values for glucose which vary from 0·4 to 140 μmol dm^{-3}.

In measuring enthalpy change in continuous culture a major and fundamental problem concerns the time lag between fermentor contents leaving the chemostat and power output being recorded. Brettel *et al* (1981b) in a study with *Sacch. cerevisiae* reduced this time to 2 min by using a high flow rate (200 cm^3 h^{-1}) to the calorimeter and returning culture fluid to the chemostat. In a study with the bacterium *Klebsiella aerogenes*, James and Djavan (1980) also achieved a minimum transit time of 2 min; the length of tubing between the chemostat and the calorimeter was kept to a minimum and the culture fluid was allowed to go to waste after passing through the calorimeter cell. However thermal measurements could not be made at $D < 0{\cdot}25$ h^{-1}.

In both of these studies the authors recognized that the power output of cells may change during transit to the calorimeter, and attempted to allow for this change. Plots of power output versus transit time showed a decline in power with increasing transit time, and heat output with zero transit time (i.e. in the fermentor vessel of the chemostat) was determined by extrapolation. In our view the necessity for extrapolation seriously limits the value of the data obtained. In general, K_{m} values for growth are small; thus, in the chemostat, equilibrium values of S will also be small (for $D = 0{\cdot}1\text{–}0{\cdot}8 \times \mu_{\text{max}}$, $S = 0{\cdot}1\text{–}4{\cdot}0 \times K_{\text{m}}$), and may therefore be subject to substantial change in transit times of only a few minutes. Even if the change in limiting substrate concentration during transit is known, the actual power output in the fermentor vessel could not be safely inferred, as changes in rates of utilization of, for example, carbon sources, will lead to altered concentrations of key metabolites with unknown effects on the activities of diverse metabolic pathways. The complexity of the situation may be illustrated by the curves of power output versus transit time which decrease linearly with *Klebsiella aerogenes* (James and Djavan 1980) and exponentially with *Sacch. cerevisiae* (Brettel *et al*

1981b). Brettel *et al* (1981b) derived a mathematical relationship between power output and transit time which supported an exponential decline; however, they assumed $S \ll K_m$, which requires some explanation.

Leiseifer and Schleser (1983) have described a calorimetric system in which transit time from the fermentor to the calorimeter cell is reduced to a few seconds. The performance of this system with chemostat continuous cultures has not been fully investigated; however, even in this system, it appears that with glucose-limited cultures of *Escherichia coli*, substantial corrections to power output need to be made to account for transit time. It is possible that for steady state chemostat cultures where K_m for the limiting substrate is relatively large, changes in substrate concentration during transit time might not be significant. Such a system could be *Sacch. cerevisiae* in glucose-limited medium in which microcalorimetric studies (Table 5.1) have indicated a K_m value for glucose of approximately 1 mmol dm^{-3}; K_m values are, typically, significantly lower than this (Pirt 1975).

The problem of transit time to the calorimeter has been overcome by the construction of a calorimeter in which the calorimeter cell also functions as a conventional fermentor (Ishikawa *et al* 1981a,b; Ishikawa and Shoda 1983; Shoda and Ishikawa 1982). This instrument has been used in batch and continuous culture studies of the growth of *E. coli* but not yet in studies of yeast. Accurate studies of enthalpy changes of cultures growing at fixed rates in constant environments (i.e. in a chemostat) may enable investigation of heats associated with the metabolism of nutrients other than the carbon source. The technique may also have application in studying the rapid effects of drugs etc. on microbial metabolism.

Despite the difficulties in accurate determination of power output in systems in which there is a transit time from the chemostat fermentor vessel to the calorimetric cell, results using such systems with yeast show many classic and thermodynamic features of yeast metabolism. These have been discussed by Lamprecht (1980) and include (*a*) the derivation of an equation for specific power output (Brettel 1974, 1977a) which comprises two terms, one dependent upon growth rate and one independent of growth rate and parallel to the requirement for maintenance energy (Pirt 1965) and (*b*) the demonstration of changes in specific power output and enthalpy change (mol glucose)$^{-1}$ at dilution rates below 0·3 h^{-1} for *Sacch. cerevisiae*, growing aerobically in a glucose-limited medium, due to repression of respiratory enzymes by high glucose concentrations (Brettel 1974, 1977a,b).

In publications since the review of Lamprecht (1980), Brettel *et al* (1981a,b,c) have presented a detailed study of the growth of *Sacch. cerevisiae* in glucose- and ethanol-limited chemostat cultures. Energy balances were determined and the proportion of energy produced and stored in cell material was determined as 45 and 60% with ethanol and glucose, respectively; the remainder of produced energy was accounted for by enthalpy changes. In ethanol medium, acetate was a presumed by-product of metabolism; it was also concluded that production of cell material from ethanol is an exothermic process.

5.2.4 Synchronous Culture Studies of the Cell Cycle

The major events of the cell division cycle and their regulation in *Sacch. cerevisiae* have been discussed by Hartwell (1974) and Hartwell *et al* (1974). The cell cycle is divided into four phases which are (in order) G1, S, G2, M. The S and M phases are those characterized by DNA synthesis and nuclear division, respectively, and G1 and G2 are the intervals between them. It has been shown by analysis of populations of *Sacch. cerevisiae* growing exponentially on various carbon sources (culture doubling times 92–435 min) that the length of the S and M phases are relatively constant (45 and 16 min, respectively); it thus appears that G1 and G2 are periods of preparation for the S and M phases (Barford and Hall 1976). Metabolic activity shows substantial variation during the cell cycle, and enzyme synthesis is essentially periodic; the synthesis of individual enzymes occurs at discrete time intervals within the cell cycle, the actual time being dependent upon the particular enzyme (Halvorsen *et al* 1971).

In yeasts, cell cycle studies have been mainly concerned with *Sacch. cerevisiae* (a budding yeast) and *Schizosacch. pombe* (a fission yeast). The number of investigations in which microcalorimetry has been used is small; these investigations have involved the growth of synchronous cultures, in which the events of the cell cycle occur in the majority of the population at the same time. The degree of synchrony in a population is often quantified using the synchrony index (F) of Blumenthal and Zahler (1962), which is given by:

$$F = \frac{N}{N_0} - 2^{t/g}$$

where N_0 and N are the cell numbers before and after division, t is the time over which cell numbers increase, and g is the generation time. In practice an index of between 0·5 and 0·7 is considered an acceptably high degree of synchrony. Sayyadi *et al* (1980) in a microcalorimetric study, used a synchrony index (S), defined by Nusse (1973), and given by:

$$S = 1 - \frac{A_{\text{ms}}}{A_{\text{es}}}$$

where, in plots of cell number versus time, A_{es} is the area between an ideal synchronous curve and the corresponding exponential curve, and A_{ms} is the area between the measured synchronous curve and the ideal synchronous curve.

Methods for obtaining synchronous cultures are diverse and often not well understood. They can be divided into selection and induction methods. In the former, cells at the same stage of the cell cycle (usually newly-formed cells) are separated from exponentially growing populations by selective centrifugation procedures, filtration, etc. In induction methods all the cells of a population are brought to the same stage of the cell cycle. The latter methods are subject to the general criticism that the behaviour of cells after induction may be a consequence of the induction procedure rather than the normal cell cycle. In synchronous cultures, the degree of synchrony declines with successive generations reflecting variation in generation time amongst the members of the population (Dean and Hinshelwood 1966).

In cultures of *Schizosacch. pombe*, synchronized by an induction method using 2-deoxyadenosine, linear increases in power output during the cell cycle were obtained (Poole 1977). Power output thus appears to be synchronized by 2-deoxyadenosine. However, the rate of oxygen uptake oscillated, so giving fluctuations in the ratio of power output to oxygen consumed. It was suggested that this fluctuation may reflect a changing efficiency of oxidative phosphorylation or a varying contribution of glycolysis to energy metabolism during the cell cycle.

In experiments with synchronous cultures of the same yeast strain, derived by a sedimentation-velocity selection method, a near-linear increase in power output was also observed in one complete cell cycle (Poole *et al* 1973). Glucose and glycine uptake also increases linearly during the cell cycle in *Schizosacch. pombe* (Kubitschek and Claymen 1976); thus a linear rather than an exponential mode of growth during the cell cycle is indicated.

The effect of the uncoupler, carbonyl-cyanide-*m*-chlorophenylhydrazone, on power output by cells derived from different stages of the cell cycle was dependent upon the method used to obtain synchrony. In selection-synchronized cells, two peaks of power were observed during the cell cycle which corresponded with peaks of oxygen uptake, whereas in 2-deoxyadenosine-synchronized cells the power increased linearly.

Poole (1977) has suggested that the accuracy afforded by continuous-flow microcalorimetric measurements of overall metabolic activity may be useful in distinguishing betwen linear and exponential growth during the cell cycle; the maximum deviation between variables that double with linear or exponential kinetics is only 6%. The results of Poole *et al* (1973) have been criticized by Newell (1980) who has suggested that linear power is a consequence of oxygen limitation.

The cell cycle of *Sacch. cerevisiae* (baker's yeast strain) has been investigated using synchronous cultures obtained by a selection method involving repeated centrifugation and filtration through a membrane filter of pore size 8 μm (Sayyadi *et al* 1980). In contrast to the results for *Schizosacch. pombe*, the power output increased exponentially for more than three generations. In this study the interaction of nystatin with cells at two different stages of the cell cycle was also followed microcalorimetrically. Cells in the G2 phase appeared markedly more sensitive than cells from an earlier stage in the cell cycle. The earlier stage was said to be the beginning of the S phase, but the timing of this was deduced from previously published work concerning the lengths of cell cycle phases, and measurements of DNA synthesis were not made.

The cell cycle of baker's yeast has also been studied in synchronous cultures obtained in a chemostat (Brettel 1977b; Brettel *et al* 1980b). In the method employed, chemostat cultures apparently came to a steady state, yet showed metabolic oscillations due to a partially synchronous growth which persisted for up to 88 generation times. Whilst, in the strict sense, if metabolic oscillations occur a steady state cannot exist (and the results show a periodic fluctuation in glucose concentration in the chemostat), the results do show periodic variation in the proportion of newly-budding cells; and periodicity was once per generation time. These oscillations were found in glucose- and ethanol-limited cultures, and in ethanol-limited cultures they could be observed after adjustment of the chemostat flow rate (Brettel *et al* 1981a). The authors of the relevant papers do not offer explanations for the achievement of synchrony.

Power output in glucose-limited 'chemostat'-synchronous cultures was markedly periodic with maxima a few minutes before budding. Oxygen consumption was also periodic with maxima occurring at the same time, but being less marked. Thus, the ratio of power output to oxygen consumption also showed a peak during the cell cycle prior to bud emergence, possibly reflecting fermentative metabolism of cell reserve material. It was suggested that the observed increase in metabolism was necessary to achieve the rate of biosynthesis required for bud emergence. Ethanol-limited 'chemostat'-synchronous cultures also showed maxima in power output, and in the ratio of power output to oxygen consumption, prior to bud emergence, although oxygen consumption was relatively stable throughout the cell cycle (Brettel 1977b). Since periodic variation in power output in 'chemostat'-synchronous cultures of baker's yeast was obtained although biomass remained constant, these results are in contrast to those obtained by the centrifugation/filtration selection method, where power output was proportional to biomass.

5.2.5 Growth Studies on Agar Media

The growth of microbial colonies on agar has been discussed by Pirt (1967, 1975). Isolated colonies of bacteria, and presumably yeast, show a linear increase in colony diameter with time. This may be explained as being due to the exponential growth of cells in a relatively thin growing zone (of fixed width) around the perimeter of the colony, while the majority of cells within the colony are non-growing. Growth of cells at the colony centre may cease as toxic products of metabolism accumulate or substrate concentrations decline, while at the colony edge growth will continue as a consequence of diffusion of nutrient from areas adjacent to the colony. Thus, growth of individual colonies on agar will be affected by the depth of agar and inoculum density. Colony height will also be limited by diffusion of substrates (including, where appropriate, oxygen) into, and of toxic products out of the colony. The maximum oxygen penetration into bacterial colonies is 30 μm (Wimpenny and Coombs 1983).

Lamprecht and Meggers (1972) followed power output during the growth of *Sacch. cerevisiae* in a calorimetric vessel lined with agar medium. Experiments were conducted aerobically and anaerobically in media with various concentrations of glucose; enthalpy changes were 1670 and 68 kJ mol^{-1} glucose, respectively, comparable with those obtained in batch culture. Power-time curves were, however, complex. In view of the heterogeneous nature of both organisms and agar during growth on solid media, and of the extra difficulties thus imposed in order to obtain analytical data, studies in liquid media would in general appear to be more suitable.

5.2.6 Studies of Non-growing Cells

The metabolism of yeast under non-growing conditions may be observed in media lacking one or more nutrients for growth. Nunomura and Fujita (1981) have investigated the endogenous metabolism of various strains of *Sacch. cerevisiae*, *Sacch. carlsbergensis* and *Candida utilis* suspended in phosphate buffer solution in the absence of an exogenous energy source. Under anaerobic conditions there was no observable power output, whereas under aerobic conditions power output was considerable; *p-t* curves showed an initial peak followed by a slow decline. Power output was accompanied by oxygen consumption, suggesting that the exothermic reaction was due to an endogenous respiration. This conclusion was supported by observations with a respiratory-deficient yeast, which gave no detectable power output under aerobic or anaerobic conditions.

Power output was strongly influenced by temperature, pH and cell history, including the nature of the culture medium used for growth and the storage conditions of cells prior to experiments. There was no detectable power evolution by starved cells which had been harvested, washed and shaken for 20 h at 30 °C in phosphate buffer. With the exception of the respiratory-deficient yeast, all strains gave essentially similar *p-t* curves, though an active-dried yeast strain gave a prolonged, high initial heat output, and *Candida utilis* gave a reduced power output compared to strains of *Saccharomyces*.

When relatively low concentrations of glucose were added to washed suspensions of *Sacch. cerevisiae* under aerobic conditions (Nunomura and Fujita 1982) *p-t* curves showed three distinct phases. In the first phase a peak was observed (at approximately 0·5 h) which was attributed to glucose uptake and was accompanied by ethanol formation. After exhaustion of glucose, the power declined to a relatively stable value which was associated with the respiration of ethanol. Upon exhaustion of ethanol (at 6 h) the power fell initially slowly, and subsequently more rapidly to undetectable levels. Power in this final phase was due to endogenous metabolism; trehalose, a reserve carbohydrate in yeast, was shown to accumulate during the first phase and be utilized in the final phase; thus either alone, or with other cell material, it was a substrate for endogenous respiration.

In these experiments, increasing glucose concentration increased peak height and width during the initial phase, and prolonged the plateau of power during the second phase. These results are thus in accord with previous studies of metabolic regulation in *Sacch. cerevisiae* which indicate that in the presence of excess glucose, respiration is inhibited in favour of fermentation. When glucose was replaced by fructose or mannose, *p-t* curves were similar to those with glucose; however, with sucrose the initial peak was absent.

Under anaerobic conditions *p-t* curves showed only the first of the phases observed under aerobic conditions, suggesting that power output in the second and third phases was due to aerobic respiration and indicating endogenous metabolism to be respiratory. Similar *p-t* curves were obtained when cells were incubated aerobically in the presence of inhibitors of respiration, or when a respiratory-deficient strain of yeast was used. By graphical integration of the area under *p-t* curves Nunomura and Fujita (1982) estimated $\Delta_{met}H$ for glucose to be 1798 and 100 kJ mol^{-1} under aerobic and anaerobic conditions, respectively. The experimental value for ethanol oxidation was 710 kJ mol^{-1} compared to a theoretical value of 1348 kJ mol^{-1}. The differences between experimental and theoretical enthalpy changes were attributed to the formation of minor metabolic products, consumption of energy for maintenance, and a lower efficiency of oxidative phosphorylation. In the study of Nunomura and Fujita (1982) results for *Sacch. carlsbergensis* and an active-dried strain of yeast were essentially similar to those of the *Sacch. cerevisiae* strain given above, although the course of trehalose accumulation differed; in particular, the initial trehalose content of the active-dried yeast was high.

Power-time curves obtained after the addition of glucose to aerobic non-growing suspensions of *Sacch. cerevisiae* have also been reported by Schaarschmidt (1972) and Lamprecht *et al* (1973). These curves show power rising to a peak after 2 and 5 h with low and high glucose concentrations, respectively. After the peak, power declined to a plateau which was maintained for more than 10 h before starting to decline. The curves are thus superficially similar to those reported by Nunomura and Fujita (1982). However, the interpretation was different. The plateau level of power output was regarded as due to the metabolism of maintenance; power evolution above the plateau was proposed to represent the deposition of reserves. Enthalpy changes associated with metabolism of maintenance and deposition of reserves were determined for various strains of *Sacch. cerevisiae* with different ploidy (i.e. number of chromosome sets), and in the presence of inhibitors. The results are fully discussed in the review of Lamprecht (1980).

Under anaerobic conditions in the presence of saturating concentrations of glucose, Beezer *et al* (1977a) observed that power output by *Sacch. cerevisiae* suspensions in buffer solution rose rapidly to a plateau value which was maintained for more than 1 h. The ability to obtain zero order kinetics is of significance to assay development (5.3.3). In this system the ability to prevent oxygen interactions (arising from oxygen leakage into the system) is of importance. In this respect the use of glucose oxidase, as an oxygen sink, or of inhibiting respiration or using a respiration-deficient mutant may be of advantage (Hoogerheide 1975).

5.3 Commercial Applications

5.3.1 Strain Identification and Characterization

The potential application of microcalorimetry to microbial identification has been investigated in many studies (e.g. Boling *et al* 1973; Staples *et al* 1973; Monk and Wadsö 1975; Ljungholm *et al* 1976; Kallings 1976). Power-time curves representing the growth of microorganisms in batch culture may show substantial detail, especially in media with more than one source of carbon and/or nitrogen, and changing oxygen tension, as in sealed-ampoule studies (Kallings 1976). The shape of the *p-t* curve will depend upon all those factors affecting growth and metabolism; it will depend upon the choice of medium, the factor(s) limiting growth, inoculum size, the length of the lag phase (which will be affected by inoculum history and possibly also inoculum size), whether or not endogenous reserves are accumulated during growth and subsequently utilized, etc. In media in which more than one source of carbon/energy is utilized it will also reflect the order and rate of adaption to utilization of these sources. A number of these factors are discussed in relation to yeast identification by Perry *et al* (1979). Furthermore, changes in the specific rate of oxygen utilization due to decreased oxygen availability will lead to marked effects on *p-t* curves, as large enthalpy changes are associated with aerobic respiration; the theoretical enthalpy change for, e.g. the ethanolic fermentation of glucose is 135 kJ mol^{-1} compared to 2872 kJ mol^{-1} for glucose respiration to carbon dioxide plus water. Inhibition of respiration in favour of fermentation by glucose, observed in glucose-sensitive yeasts, would also markedly affect *p-t* curves.

The influence of physical parameters, particularly of oxygen tension, on *p-t* curves of *Sacch. cerevisiae* and *Sacch. fragilis* has been discussed by Beezer *et al* (1979).

To obtain good reproducibility in *p-t* curves, for a single microbial strain, inocula must be grown under standardized conditions. The use of inocula derived from a single growth of cells and stored in liquid nitrogen (Beezer *et al* 1976) gave *p-t* curves in which the maximum variation in replicate experiments was ±5% (Perry 1980). In experiments using inocula derived from different growths of the same yeast strain, Perry *et al* (1979) found good reproducibility between *p-t* curves when cells were used at the same inoculum density and were grown in defined, minimal medium after a growth regimen involving a number of subcultures (in batch culture). Batches of cells derived from different growths in a complex medium (modified antibiotic medium No. 3) gave *p-t* curves with a poor reproducibility. Thus, proposed microcalorimetric identification procedures involving inocula grown in complex media, or in some cases the direct use as inocula of organisms in patient's blood or cerebrospinal fluid (Kallings 1976) will require extensive validation. A further area of difficulty concerns the nature of the medium chosen for calorimetric incubations; *p-t* curves obtained using media containing undefined components such as tryptose, yeast extract, etc. are unlikely to give sufficiently high reproducibility for identification studies (e.g. Fujita *et al* 1978).

A more fundamental difficulty in the microcalorimetric identification of microorganisms is that microbial strains similar in many features of cell structure and metabolism may, due to relatively small differences in metabolic regulation or the ability to use a particular substrate, give vastly different *p-t* curves. In a flow microcalorimetric study Perry *et al* (1983c) found marked differences in the *p-t* curves of brewing, baking and distilling strains of *Sacch. cerevisiae* in defined media containing glucose, glucose plus maltose or maltose as the carbon source (Figure 5.1); similarly, in the same study, a brewing strain of *Sacch. uvarum* and a mutant strain derived from it in a brewing process gave substantially different *p-t* curves in a medium containing glucose. Also, on medium containing glucose plus maltose, Schaarschmidt and Lamprecht (1978) reported a biphasic *p-t* curve for a wild yeast strain, whereas Perry *et al* (1983b) observed a monophasic response with several commercial yeast strains.

It seems unlikely, therefore, that a single microcalorimetric procedure suitable for the identification of a wide range of microorganisms could be found. However, in the flow microcalorimetric study of Perry *et al* (1983c) referred to above, characterization of representative strains of *Sacch. cerevisiae* and *Sacch. uvarum* used in the food industry was achieved, and a microcalorimetric procedure was proposed as a rapid alternative to

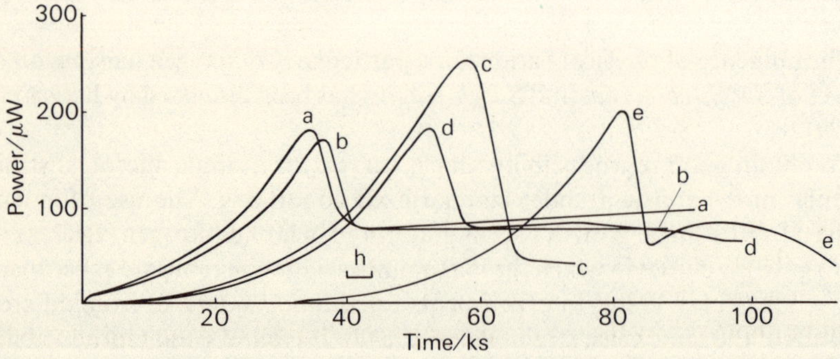

Figure 5.1. *p-t* curves showing the growth of *Sacch. cerevisiae*: a, pressed baking strain; b, active dried baking strain; c,d, distilling strains; e, brewing strain. Growth medium contained 10 mmol dm^{-3} maltose (after Perry *et al* 1983c).

conventional methods of characterization of commercial yeast strains. In a study of lactic acid bacteria used in the dairy industry, Fujita *et al* (1978) similarly proposed a flow microcalorimetric method for the rapid identification of strains suitable for commercial use. They also used a defined medium (containing several carbohydrates), and noted that satisfactory reproducibility of *p-t* curves was not obtained in media containing undefined components (e.g. peptone), which were subject to batch-to-batch variation.

5.3.2 Evaluation of Complex Media

Commercial substrates for the growth of yeasts are often industrial by-products which may show substantial variability. Molasses, a waste product of the sugar industry, has found application as a raw material for the production of baker's yeast and ethanol (Olbrich 1963; Ruter 1975; Sobkowicz 1976). The suitability of molasses for yeast growth depends primarily on the technical processes by which the sugar is manufactured, although agronomic factors and treatments of beet or cane are also of significance. Evaluation of molasses by physical and chemical means requires extensive work and most laboratories have developed yeast growth or fermentation methods (Baker 1979; Labendzinski 1979; Oura 1979). Such methods are, however, time-consuming and as they rely on end-point analysis they do not provide a continuous record of yeast activity.

Yeast growth may be readily monitored by flow microcalorimetry and Perry *et al* (1981) have demonstrated the applicability of the technique to the comparison of batches of molasses. They used two strains of yeast (a pressed baker's strain and a distillery strain), inocula of each of which were derived from a single growth and stored in liquid nitrogen to ensure reproducibility, and six samples of molasses; five of these were identified industrially as being satisfactory and one as being poor. Calorimetric incubations were conducted anaerobically in a medium containing molasses, ammonium and potassium phosphates, and antibacterial antibiotics. Power-time curves were reproducible (Figure 5.2), but were remarkably different for the two strains of yeast used; with industrially acceptable molasses the values of p_{max} were lower, and the times required to reach p_{max} were longer with the baking than with the distilling strain. However, the industrially poor molasses also gave *p-t* curves substantially different from those of acceptable batches (Figure 5.2); with industrially poor molasses and the distilling yeast strain, p_{max} was halved, and the time to reach p_{max} doubled; with the baking strain p_{max} was increased threefold. It was concluded that the microcalorimetric technique was applicable to determinations of the suitability of batches of molasses for yeast growth and/or ethanol production, and that discrimination between batches might be improved by the use of different yeast strains, or strains grown under different physiological conditions. It is probable that this technique could be applied to the evaluation of components of other complex media. In this respect, an advantage quoted for the technique is that homogeneous solutions

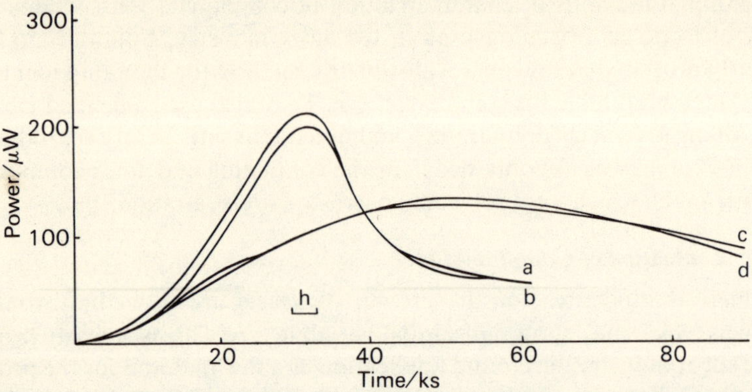

Figure 5.2. *p-t* curves showing the growth of a distillery strain of *Sacch. cerevisiae* in an industrially acceptable molasses sample (a,b) and in an industrially poor molasses sample (c,d). Stock molasses solution prepared independently for replicates (after Perry *et al* 1981).

of substrates are not required, and suspensions could be used provided any particles present would pass through the tubing of the flow calorimeter.

5.3.3 Assay of Antifungal Compounds and Studies of Their Interaction with Yeast

It is clear from the preceding sections of this review that microcalorimetry may be used as a sensitive technique to follow microbial growth and/or metabolism. Antimicrobial agents would, therefore, be expected to, and have been demonstrated to, have marked effects on *p-t* curves. Prat (1953) demonstrated that streptomycin affected the *p-t* curve of *E. coli* cultures and there have been many subsequent reports of a qualitative or semi-quantitative nature concerning antibiotic/cell interactions (Beezer and Chowdhry 1980a). However, in a small number of studies quantitative assays have been developed. Joslin (1984) has described flow microcalorimetric assays for the antibacterial antibiotics zinc bacitracin, polymyxin B sulphate and neomycin sulphate using growing cultures of *Micrococcus luteus*, *Bordatella bronchiseptica* and *Bacillus pumilis*, respectively, and less sensitive but more rapid assays for polymyxin B sulphate and zinc bacitracin using respiring (non-growing) cells of *E. coli*. Perry *et al* (1980) have also described a flow microcalorimetric assay for phenol using respiring cells of *E. coli*. In these interactions power was not always inhibited by the antimicrobial compound; low phenol concentrations stimulated power output, as did neomycin sulphate when added to growing cells.

With regard to yeast studies, Beezer *et al* (1977a) developed a flow microcalorimetric assay for the antifungal antibiotic nystatin in which *Sacch. cerevisiae* NCYC 239 was the test organism. This strain was chosen as it did not aggregate under the experimental conditions employed. Reproducibility in the assay was dependent upon the use of a standard inoculum obtained by using cells from a single growth stored in liquid nitrogen (Beezer *et al* 1976); similarly prepared inocula were also used in the assay of antibacterial antibiotics.

In the nystatin assay, yeast was inoculated into phthalate buffer solution containing glucose (10 mmol dm^{-3}) which was maintained at 25 °C; anaerobic conditions were achieved by flushing the medium with nitrogen before and during incubation. In the absence of antibiotic the power rose to a plateau value which was maintained for at least 1 h. In medium containing nystatin, however, power rose initially and then declined, the rate of decline increasing with increasing nystatin concentration. The decay curves did not show simple first or second order kinetics; thus a single rate parameter related to nystatin concentration could not be derived. However, there was a linear relationship between log [nystatin] and the time required for power to fall to certain values. Compared to conventional microbiological assays for nystatin (e.g. agar-plate diffusion assay) the microcalorimetric assay had better reproducibility ($\pm 3\%$ compared to $\pm 10\%$ for the plate assay), was more sensitive, required less time (only 1 h) and was more adaptable to automation.

Similar assays using non-growing suspensions of *Sacch. cerevisiae* have been developed for the polyene antibiotics nystatin, candicidin, amphotericin B, filipin and lucensomycin (Beezer *et al* 1977b) and N-acetylnystatin (Beezer and Chowdhry 1980b). In these assays, performed at 30 °C, a linear relationship was obtained between antibiotic concentration and the time required for power to fall to a specified value. The empirical nature of the assays derived is emphasized by the different type of response to nystatin at 30 °C compared to that at 25 °C. The nystatin assay was used to compare the activity of various formulations of nystatin complexed with polyvinylpyrrolidone.

Flow microcalorimetric assays for polyene antibiotics (nystatin, amphotericin B, candicidin, pimaricin and lucensomycin) have also been developed which are based on the addition of antibiotic to exponentially growing cultures of *Sacch. cerevisiae* (Beezer and Chowdhry 1981). High antibiotic concentrations caused a decline in power, whereas lower concentrations caused an initial decline followed by a rise in power, but at a lower rate than observed in control cultures without antibiotic. Assays were based on the observed linear relationship between antibiotic concentration and initial decrease in power, and were reproducible ($\pm 2{\cdot}4\%$) and sensitive (lowest detectable concentration 1 μmol dm^{-3}); assay time required was 9·5 h. It was noted in this study that the ranking order of the polyenes investigated, as regards ability to inhibit power output, was different to that generally accepted for membrane activity or ability to prevent yeast growth.

Flow microcalorimetry has also been applied to investigations of the antimicrobial effects of drug combinations. Cosgrove *et al* (1978) determined the minimum inhibitory concentration, and effect on power output and cytoplasmic leakage, of various combinations of amphotericin B with the relatively new imidazole antifungal compounds clotrimazole and miconazole. Definite antagonistic responses were obtained. In comparing the methods used the authors concluded that effects on power output were of more significance than measurements of leakage of individual cell components (e.g. K^+, Mg^{++}). Microcalorimetry has also been used to study the effect of antibiotic combinations on bacteria (Joslin 1984).

In addition to the application of flow microcalorimetry to drug assay, it has been suggested (Beezer and Chowdhry 1980a) that the technique might be used to investigate the mode of action and design of new drugs. There are several interesting observations in the literature (see review of Beezer and Chowdhry 1980a); for example, lucensomycin gave two types of response when added to respiring yeast cells. At low concentrations ($< 5\,\mu\text{mol dm}^{-3}$) a plateau of power was achieved which was lower than that of control cells, whereas above $5\,\mu\text{mol dm}^{-3}$ the power declined in a manner similar to that for nystatin. It has been suggested (Beezer *et al* 1977b) that this reflects differing, concentration dependent, modes of action.

The effect of various physical and chemical factors on the interaction of polyene antibiotics with respiring cells of *Sacch. cerevisiae* has also been studied by flow microcalorimetry (Beezer and Chowdhry 1980c). The bioactivity of nystatin decreased as the pH of the medium increased towards neutral, and calcium and magnesium ions were shown to attenuate the effect of nystatin, amphotericin B and candicidin, but not of filipin or pimaricin. Polyene 'complexed' with ergosterol or cholesterol had reduced activity, except that the activity of pimaricin was unaltered, perhaps indicating a weaker polyene-sterol interaction.

5.4 Concluding Remarks

There are many studies which demonstrate the ability of microcalorimetric procedures to detect yeast growth or metabolism. Microcalorimetry has, therefore, an immediate and potentially important application in monitoring industrial fermentations. The advantages of microcalorimetry in this context are that measurements are reproducible, constantly available and can be readily related to biomass or rates of metabolism. Unlike many other techniques it is not dependent upon having homogenous media.

There would also seem to be a considerable potential for microcalorimetric assays. The advantages of the polyene assays described here, compared to conventional assays, are speed, sensitivity and reproducibility. There is no apparent reason why assays with similar advantages should not be developed for other antifungal antibiotics. Calorimetry also offers the potential to study synergism and antagonism in drug combinations, and, possibly, in a more meaningful way than determining minimum inhibitory concentrations. The application of microcalorimetry to evaluation of variable complex commercial substrates (e.g. molasses) has also been proposed, and the technique might have potential in microbiological assays of vitamins, amino acids, etc.

The principle of identification of microorganisms by analysis of their *p-t* curves in suitable media has been widely supported. However, the results obtained with yeasts suggest that strains belonging to the same species may give such different *p-t* curves that this is not possible. Conversely, microcalorimetry may thus be used to differentiate between closely related strains, and results reviewed here suggest that by careful choice of inoculum density and medium composition, differentiation of the major groups of commercial yeast strains is possible.

Microcalorimetry has been used in many studies to obtain kinetic and thermodynamic data of yeast growth and metabolism. The technique has been used to determine K_m and V_{max} values for sugar utilization during growth (Table 5.1), and to obtain thermodynamic data for sugar uptake processes. Enthalpy changes have been derived for growth and metabolism in batch culture; however, values derived may show significant differences from theoretical values for the utilization of the 'energy' source. It is not clear whether these differences relate to experimental errors or to reactions not involving the 'energy' source directly. The most suitable system in which to determine energy balances is chemostat continuous culture. The chemostat yeast studies so far reported may be criticized in that, in order to measure power output, samples were removed from the chemostat, thus destroying the steady state. The problem has been recently overcome by the development of an instrument in which the fermentor vessel functions also as the calorimeter vessel (Ishikawa and Shoda 1983). This instrument might usefully be applied to yeast studies.

There have been few microcalorimetric synchronous culture studies. The interaction of yeast with substrates, antibiotics, preservatives, etc. might be expected to vary during the cell cycle. Microcalorimetry offers a rapid means of studying interactions. The observation of Sayyadi *et al* (1980) that the variation in nystatin susceptibility of yeast cells during the cell cycle may be monitored microcalorimetrically indicates the potential for further studies.

Finally, perhaps the most surprising feature of microcalorimetric yeast studies is the range of organisms which have been used. The yeasts are a diverse group. Yet the majority of studies have used only strains of *Sacch. cerevisiae* and closely related species. In particular, there have been few studies of *Candida* sp. which are used as food yeasts and are industrially important as they may be grown on hydrocarbons and industrial by-products, including molasses, wood hydrolysates and residual liquors from food and chemical processing (Peppler 1970).

References

Baker B F (1979) In: Sinda E, Parkkinen E (eds) *Proceedings of Symposium*: *Problems with Molasses in the Yeast Industry*. Helsinki, Finnish State Alcohol Company, pp 113–27

Barford J P and Hall R J (1976) *Exp Cell Res* **102**, 276–84

Battley E H (1960) *Physiologia Pl* **13**, 628–40

Beezer A E and Chowdhry B Z (1980a) In: Beezer A E *Biological Microcalorimetry*. London, Academic Press, pp 195–246

Beezer A E and Chowdhry B Z (1980b) *Talanta* **27**, 1–5

Beezer A E and Chowdhry B Z (1980c) *Microbios* **28**, 107–21

Beezer A E and Chowdhry B Z (1981) *Experientia* **37**, 828–30

Beezer A E, Newell R D and Tyrrell H J V (1976) *J Appl Bacteriol* **41**, 197–207

Beezer A E, Newell R D and Tyrrell H J V (1977a) *Anal Chem* **49**, 34–7

Beezer A E, Chowdhry B Z, Newell R D and Tyrrell H J V (1977b) *Anal Chem* **49**, 1781–4

Beezer A E, Newell R D and Tyrrell H J V (1979) *Antonie van Leeuwenhoek* **45**, 55–63

Belaich J-P, Senez J C and Murgier M (1968) *J Bacteriol* **95**, 1750–57

Belaich J-P, Beezer A E, Prosen E and Wadsö I (1981) *Codata Bulletin* **44**, 4–10

Blumenthal L K and Zahler S A (1962) *Science* **135**, 724

Boling E A, Blanchard G C and Russell W J (1973) *Nature* **241**, 472–3

Brandli E (1980) *Thesis*. Eidgenossische Technische Hochschule, Zurich

Brettel R (1974) In: Klaushofer H, Sleyter U B (eds) *Proceedings of the Fourth International Symposium on Yeasts, Vienna*. Wien, University of Agriculture, part 1,B6, pp 87–8

Brettel R (1977a) *Mikrocalorimetrische Untersuchungen an Kontinuierlichen und Statischen Kulturen der Hefe Saccharomyces cerevisiae*. Thesis, Free University of Berlin

Brettel R (1977b) In: Lamprecht I and Schaarschmidt B (eds) *Application of Calorimetry in the Life Sciences*. Berlin, De Gruyter, pp 129–38

Brettel R, Lamprecht I and Schaarschmidt B (1980a) *Radiat Environ Biophys* **18**, 301–9

Brettel R, Lamprecht I, Sayyadi P and Schaarschmidt B (1980b) In: *Proceedings of the Sixth International Conference on Thermal Analysis, Bayreuth, Federal Republic of Germany*. Basel, Birkhäuser Verlag, pp 553–7

Brettel R, Lamprecht I and Schaarschmidt B (1981a) *Eur J Appl Microbiol Biotechnol* **11**, 212–5

Brettel R, Lamprecht I and Schaarschmidt B (1981b) *Eur J Appl Microbiol Biotechnol* **11**, 205–11

Brettel R, Lamprecht I and Schaarschmidt B (1981c) *Thermochim Acta* **49**, 53–61

Cosgrove R F, Beezer A E and Miles R J (1978) *J Infect Dis* **138**, 681–5

David M H (1974) *J Inst Brew* **80**, 80–5

Dawes I W (1976) In: Skinner F H, Hugo W B (eds) *Inhibition and Inactivation of Vegetative Microbes*. Soc Appl Bacteriol Symp Series No. 5, London and New York, Academic Press, pp 279–304

Dean A C R and Hinshelwood C N (1966) *Growth, Function and Regulation in Bacterial Cells*. London, Oxford University Press

Eddy A A (1982) In: Rose A H, Morris J G (eds) *Advances in Microbial Physiology*, vol 23. London and New York, Academic Press, pp 1–78

Fiechter A, Fuhrmann G F and Kappeli O (1981) In: Rose A H, Morris J G (eds) *Advances in Microbial Physiology*, vol 22. London and New York, Academic Press, pp 123–84

Fujita T, Monk P R and Wadsö I (1978) *J Dairy Res* **45**, 457–63

Gottlieb S F (1971) *Ann Rev Microbiol* **25**, 111–52

Gustafsson L (1979) *Arch Microbiol* **120**, 15–23

Gustafsson L and Norkrans B (1976) *Arch Microbiol* **110**, 177–83

Halvorsen H O, Carter B L A and Tauro P (1971) In: Rose A H, Morris J G (eds) *Advances in Microbial Physiology*, vol 6. London and New York, Academic Press, pp 47–106

Hartwell L H (1974) *Bacteriol Rev* **38**, 164–94

Hartwell L H, Culloti J, Pringle J R and Reid B J (1974) *Science* **183**, 46–51

Hashimoto M and Takahashi K (1982) *Agric Biol Chem* **46**, 1559–64

Hoogerheide J C (1975) *Radiat Environ Biophys* **11**, 295–307

Ishikawa Y, Shoda M and Maruyama H (1981a) *Biotechnol Bioeng* **23**, 2629–40

Ishikawa Y, Nonoyama Y and Shoda M (1981b) *Biotechnol Bioeng* **23**, 2825–36

Ishikawa Y and Shoda M (1983) *Biotechnol Bioeng* **25**, 1817–27

Itoh S and Takahashi K (1984) *Agric Biol Chem* **48**, 271–5

James A M and Djavan A (1980) *Microbios* **29**, 171–83

Joslin N (1984) *Bioassay and Interactions of Antibiotics with Sensitive Cells. A Microcalorimetric Study*. Ph.D Thesis, University of London

Kallings L O (1976) In: Johnston H H, Newsom S W B (eds) *2nd International Symposium on Rapid Methods and Automation in Microbiology*. Oxford and New York, Learned Information Europe Ltd, pp 140–3

Kreger-van Rij N J W (1969) In: Rose A H, Harrison J H (eds) *The Yeasts*, vol 1. London and New York, Academic Press, pp 5–78

Kubitschek H E and Claymen R V (1976) *J Bacteriol* **127**, 109–13

Labendzinski S (1979) In: Sinda E, Parkkinen E (eds) *Proceedings of Symposium: Problems with Molasses in the Yeast Industry*. Helsinki, Finnish State Alcohol Company, pp 39–56

Lamprecht I (1969) *Mikrokalorimetrische Untersuchungen an Hefenkulturen und Beiträge zur theoretischen Deutung*. Thesis, Free University of Berlin

Lamprecht I (1980) In: Beezer A E (ed) *Biological Microcalorimetry*. London and New York, Academic Press, pp 43–112

Lamprecht I and Meggers C (1969) *Z Naturforsch* [*C*] **24b**, 1205–7

Lamprecht I and Meggers C (1972) *Biophysik* **8**, 316–25

Lamprecht I and Schaarschmidt B (1973) *Bull Soc Chim Fr* **4**, 1200–1

Lamprecht I, Meggers C and Stein W (1971) *Biophysik* **8**, 42–5

Lamprecht I, Schaarschmidt B and Stein W (1973) *Biophysik* **10**, 177–86

Leiseifer H P and Schleser G H (1983) *Z Naturforsch* [*C*] **38c**, 259–67

Ljungholm K, Wadsö I, and Mårdh P-A (1976) *J Gen Microbiol* **96**, 283–8

Loureiro-Dias M C and Arrabaca J D (1982) *Z Allg Mikrobiol* **22**, 119–22

Matile Ph, Moor H and Robinow C F (1969) In: Rose A H, Harrison J H (eds) *The Yeasts*, vol 1. London and New York, Academic Press, pp 219–302

Monk P and Wadsö I (1975) *J Appl Bacteriol* **38**, 71–4

Morris J G (1975) In: Rose A H, Morris J G (eds) *Advances in Microbiol Physiology*, vol 12. London and New York, Academic Press, pp 169–246

Murgier M and Belaich J-P (1971) *J Bacteriol* **105**, 573–9

Newell R D (1975) *Bioassay of Polyene Antibiotics by Flow Calorimetry*. Ph.D Thesis, University of London

Newell R D (1980) In: Beezer A E (ed) *Biological Microcalorimetry*. London and New York, Academic Press, pp 163–86

Nichols S C, Prichard F E and James A M (1979) *Microbios* **25**, 187–203

Nunomura K and Fujita T (1981) *J Gen Appl Microbiol* **27**, 357–66

Nunomura K and Fujita T (1982) *J Gen Appl Microbiol* **28**, 479–90

Nusse M (1973) Thesis, J.W. Goethe Universitat, Frankfurt am Main

Olbrich H (1963) In: Honig P (ed) *Principles of Sugar Technology*, vol 3. Amsterdam, Elsevier, pp 511–696

Oura E (1979) In: Sinda E, Parkkinen E (eds) *Proceedings of Symposium: Problems with Molasses in the Yeast Industry*. Helsinki, Finnish State Alcohol Company, pp 77–80

Peppler H J (1970) In: Rose A H, Harrison J H (eds) *The Yeasts*, vol 3. London and New York, Academic Press, pp 421–62

Perry B F (1980) *Flow Microcalorimetric Studies of Some Commercial Yeasts. Storage, Metabolism & Growth*. Ph.D Thesis, University of London

Perry B F, Beezer A E and Miles R J (1979) *J Appl Bacteriol* **47**, 527–37

Perry B F, Beezer A E and Miles R J (1980) *Microbios* **29**, 81–7

Perry B F, Beezer A E and Miles R J (1981) *Microbios* **32**, 163–72

Perry B F, Beezer A E and Miles R J (1983a) *Microbios* **37**, 117–29

Perry B F, Beezer A E and Miles R J (1983b) *Microbios Lett* **24**, 121–7

Perry B F, Beezer A E and Miles R J (1983c) *J Appl Bacteriol* **54**, 183–9

Pirt S J (1965) *Proc R Soc Lond [Biol]* **163**, 224–31

Pirt S J (1967) *J Gen Microbiol* **47**, 181–9

Pirt S J (1975) *Principles of Microbe and Cell Cultivation*. London and New York, Academic Press

Poole R K (1977) *J Gen Microbiol* **103**, 19–27

Poole R K, Lloyd D and Kemp R B (1973) *J Gen Microbiol* **77**, 209–20

Prat H (1953) *Rev Can Biol* **12**, 19–34

Rose A H and Harrison J H (eds) (1971) *The Yeasts*, vol. 2. London and New York, Academic Press, pp 1–2

Ruter P (1975) *Agriculture Services Bulletin Number* 25. Rome, FAO

Sayyadi P, Lamprecht I and Schaarschmidt B (1980) *Thermochim Acta* **40**, 63–71

Schaarschmidt B (1972) *Mikrokalorimetrische Untersuchungen zum Betreibs-stoffwechsel UV-bestrahlten Hefemutanten*. Thesis, Free University of Berlin

Schaarschmidt B and Lamprecht I (1977) *Radiat Environ Biophys* **14**, 153–60

Schaarschmidt B and Lamprecht I (1978) *Thermochim Acta* **22**, 333–8

Schaarschmidt B, Lamprecht I and Welge G (1974) *Radiat Environ Biophys* **11**, 53–61

Schaarschmidt B, Zotin A, Brettel R and Lamprecht I (1975) *Arch Mikrobiol* **105**, 13–16

Shoda M and Ishikawa Y (1982) In: Holme A (ed) *Modelling and Control of Biotechnical Processes; Proceedings of the First IFAC Workshop, Helsinki, Finland*. Oxford, Pergamon Press, pp 33–41

Sobkowicz G (1976) In: Birch G G, Parker K J and Wogan J T (eds) *Food from Waste*. London, Applied Science, pp 42–57

Staples B R, Prosen E J and Goldberg R N (1973) National Bureau of Standards, NBSIR 73-181, Washington DC

Suomalainen H and Oura E (1971) In: Rose A H, Harrison J H (eds) *The Yeasts*, vol 2. London and New York, Academic Press, pp 3–74

Wadsö I (1980) In: Beezer A E (ed) *Biological Microcalorimetry*. London and New York, Academic Press, pp 247–74

Walker H W and Ayres J C (1970) In: Rose A H, Harrison J H (eds) *The Yeasts*, vol 3. London and New York, Academic Press, pp 464–528

Wickerham L J (1951) *Technical Bulletin No*. 1069. U S Dept of Agriculture, Washington DC

Wimpenny J W T and Coombs J P (1983) *J Gen Microbiol* **129**, 1239–42

Chapter 6

In Vitro Thermal Studies of Blood Cells

M. Monti

6.1 Introduction

The continuous development of highly sensitive microcalorimeters in the last 20 years has made possible the measurement of heat production rate in relatively small samples of blood cells. Metabolic activity in isolated cells is usually assessed by measurement of oxygen consumption. Calorimetry offers a new approach to the estimation of total metabolism that also takes into account the anaerobic processes. In fact, anaerobic metabolism can be by far the most important metabolic activity for some cells, as in the erythrocytes where more than 90% of the substrate is broken down anaerobically. By stimulation or inhibition of some metabolic pathways, more specific information can be obtained from the measurement of heat production.

The present review covers calorimetric studies on human blood cells, concerning the development of measurement methods, the collection of data from normal subjects, and clinical applications in different diseases. Since the previous reviews in the field (Monti and Wadsö 1979; Levin 1980), several studies have been published, in particular the extension of calorimetric investigations to white blood cells (lymphocytes, granulocytes) and platelets.

6.2 Calorimeters

All calorimetric measurements on blood cells reviewed in the present chapter have been performed with heat conduction calorimeters. Static 'ampoule' and flow calorimeters have been used for most of the investigations. In recent years a prototype of the new multichannel ampoule calorimeter (Suurkuusk and Wadsö 1982) has been used by Monti and his group in several studies.

'Perfusion' (Wadsö 1974) and batch calorimeters have also been employed in a few investigations. More detailed aspects of calorimetric instruments and technique are described elsewhere (Chapter 3).

6.3 Erythrocytes

The erythrocytes, or red blood cells, represent the largest cell population in the blood, $4\text{–}6 \times 10^{12}$ cells dm^{-3}. Their function is to supply oxygen to the tissues.

6.3.1 Methodology

6.3.1.1 Preparation Procedures

Venous blood is collected in tubes containing an anticoagulant, usually heparin. Erythrocytes are separated from other blood cells by repeated centrifugations and washings or by the gel adsorption technique according to Nakao *et al* (1973). The latter preparation method is to be preferred for several reasons, the most important being the fact that by repeated centrifugations a significant fraction of the younger erythrocytes is removed. A higher rate of heat production or power (p) was in fact found in cell suspensions prepared by the gel method (Monti and Wadsö 1976d), presumably due to the presence of younger cells, expected to be metabolically more active than older erythrocytes.

6.3.1.2 Suspension Medium

Calorimetric measurements have been performed on erythrocytes suspended in different media: plasma, serum and glucose phosphate buffer solution. The importance of suspension medium was studied by simultaneous measurements of the heat production of erythrocytes in identical static ampoule calorimeters. Significantly higher (p) values were found in buffer suspensions, presumably due to the stimulating effect of phosphate on erythrocyte glycolysis (Monti and Wadsö 1976d). De Verdier (1981) also found variations in heat production, besides ATP consumption and lactate formation, when erythrocytes were suspended in different media.

6.3.1.3 Variation of pH and Temperature

The heat production rate increases linearly with increase in pH in the physiological range (Levin 1973b; Monti and Wadsö 1976c); 1·2% if the pH is increased by 0·01 units (Monti and Wadsö 1976c). A similar effect was found by increasing temperature, in the range of 32–42 °C. The temperature coefficient Q_{10} was 2·8 (Monti and Wadsö 1976c), corresponding to a 10% increase in heat production rate per degree at 37 °C.

6.3.1.4 Variation in Glucose Concentration, Storage and Calorimetric Technique

Variations in glucose concentration in the medium, in the range 3–32 mmol dm^{-3}, did not affect the heat production rate (Monti and Wadsö 1976c). Slight but significant variations in *p* value were found following storage for a few hours after cell separation (Monti and Wadsö 1976c). No significant variations of *p* values were found when measurements were performed simultaneously using different calorimetric techniques (Monti and Wadsö 1976d). Ikomi-Kumm (1977) has investigated the influence of cellular flow and viscosity on erythrocyte heat production rate. Flow rates above 25 cm^3 h^{-1} resulted in unstable power-time curves, therefore less reliable *p* values. In addition, variations of erythrocyte viscosity may introduce variable systematic errors due to heat of friction when measurements are performed with flow calorimeters. Jones and Zala (1974), using a batch calorimeter, measured heats of dilution of erythrocytes in buffer solutions.

6.3.2 Basal Heat Production

Basal erythrocyte heat production rate has been found by several investigators (Levin and Boyo 1971; Boyo and Ikomi-Kumm 1972; Levin 1973b; Monti and Wadsö 1973, 1976d) to be about 100 mW dm^{-3} packed cells (Table 6.1), corresponding to a *p* value of about 10 fW cell^{-1}. Small variations are apparently due to different experimental conditions, as methodological studies have shown that variations of pH, cell contamination, temperature, storage condition, method of preparation and suspension media may have significant effects on the results. It is therefore necessary to strictly define the experimental conditions used. Results obtained from different investigators must, when necessary, be corrected to defined standard conditions. No difference in heat production rate was found in erythrocytes from two different populations, Nigerians and Swedes (Ikomi-Kumm and Boyo 1977).

Attempts have been made by several investigators to identify the origin of the heat in erythrocyte suspensions by relating it to enthalpy changes expected

Table 6.1 Heat production rate in erythrocytes from normal human subjects

Calorimeter	*Heat production rate* /mW dm^{-3} $\pm$ SD	*Source*
flow	115 ± 19	Levin and Boyo (1971)
flow	100 ± 13	Boyo and Ikomi-Kumm (1972)
flow	116 ± 18	Levin (1973b)
static ampoule	80 ± 6	Monti and Wadsö (1973)[a]
static ampoule	100 ± 10	Monti and Wadsö (1976d)[b]

[a]Cell preparation by repeated centrifugation. The lower value of heat production rate is presumably due to the removal of younger erythrocytes.
[b]Cell preparation by gel adsorption technique.

from different metabolic pathways (Monti and Wadsö 1973; Levin *et al* 1974; Minakami and de Verdier 1976; Grarup *et al* 1977; Ikomi-Kumm 1984). Sodium fluoride, an inhibitor of glycolysis, reduces erythrocyte heat production rate by about 50%. It is, however, still unclear whether other unknown metabolic processes, besides aerobic metabolism, account for the residual heat.

6.3.3 Stimulated Erythrocytes

Stimulation of the pentose phosphate shunt in erythrocytes by methylene blue, an electron acceptor, has been found to be accompanied by a large increase in the heat production rate. Levin (1973a) has obtained p values about 4 times the basal heat production when erythrocytes were exposed to methylene blue. Monti and Wadsö (1976b) using optimal concentrations of methylene blue obtained p values about 8–10 times the basal heat production. Minakami and de Verdier (1976) have observed that the increase in heat production rate in erythrocytes stimulated with methylene blue correlated with the oxygen uptake. Methylene blue was also used by Monti to stimulate erythrocytes from patients with kidney insufficiency (1977a), and liver diseases (1977b). (See also 6.3.4.)

6.3.4 Clinical Studies

6.3.4.1 Anaemia

Boyo and Ikomi-Kumm (1972) have found increased heat production rate in erythrocytes from patients with sickle cell anaemia, while Levin and Boyo (1971) as well as Monti and Wadsö (1973) and Levin (1973b) have recorded increased values in other types of anaemia. Cerny *et al* (1974) have studied, by microcalorimetry, changs of erythrocyte shape in patients with sickle cell disease. The high values of erythrocyte heat production rate in different types of anaemia might be due to the presence of a younger cell population, with high metabolic activity, taking into account that the production of erythrocytes in the bone marrow increases in order to compensate for the anaemia. The calorimetric results are in agreement with the observation that anaemia is accompanied by increased erythrocyte glucose consumption (Brewer 1970). The rate of heat production by erythrocytes has also been studied in anaemic patients with renal failure (Monti 1977a) and patients with liver diseases (Monti 1977b). High values were found when the cells were in a resting state, whereas during stimulation of the pentose shunt with methylene blue, increased values were found in patients with renal failure and normal values in patients with liver diseases. These results give useful information concerning the mechanism by which anaemia develops during renal and liver failure. In fact, they do not give support to the hypothesis that anaemia in these patients is due to decreased aerobic metabolism by the erythrocytes.

Table 6.2 Heat production rate in erythrocytes from hyperthyroid patients before and after treatment, as well as in erythrocytes from healthy subjects

	Heat production rate/mW dm^{-3} ± SD		
	Hyperthyroid		*Healthy subjects*
Suspension medium	*Before treatment*	*After treatment*	
plasma	102 ± 15	85 ± 14	78 ± 5
buffer solution	111 ± 20	95 ± 12	86 ± 8

6.3.4.2 *Thyroid Diseases*

The heat production rate of erythrocytes from hyperthyroid patients suspended in autochthonous plasma or buffer medium was measured (Monti and Wadsö 1976a). Significantly increased *p* values were found before treatment, when comparison was made with corresponding values for normal subjects (Table 6.2). After treatment the *p* values returned to normal. A good correlation was found between the individual heat production rate and the clinical condition of the patients. Calorimetric measurements thus seem to provide useful information about the peripheral effect of thyroid hormones. Furthermore, calorimetry offers a new possibility to study the mechanism of action of thyroid hormones on the cell calorigenesis, still incompletely understood. The Na-K pump in the cell membrane has been proposed to be the site of metabolic stimulation by the thyroid hormones, leading to increased energy expenditure. In a recent study Monti *et al* (1986c) have investigated the effect on erythrocyte heat production rate by inhibition of Na-K ATPase with ouabain in hyperthyroid patients before and after treatment. Normal values, without variation during treatment, were obtained. The increased erythrocyte energy expenditure in hyperthyroidism is apparently not due to stimulation of the Na-K pump by the thyroid hormones.

6.3.4.3 *Obesity*

A number of investigators have shown results of studies that give support to the hypothesis that variations in the cell's metabolic efficiency may play an important role in the development of obesity. In recent years several investigators have claimed that Na-K pump activity is reduced and, as a consequence, the resulting disturbance in cellular thermogenesis leads to obesity. Monti and Ikomi-Kumm (1985) have quantified the change in cell metabolism induced by specific inhibition with ouabain of the Na-K pump in erythrocytes of normal and obese subjects. The decrease in *p* induced by ouabain was the same in the two groups, thus arguing against the hypothesis that a metabolic derangement secondary to defective Na-K pump activity is of importance in the development of obesity.

Binford *et al* (1973) have found a significant correlation between heat production and the quantity of *Prymnesium parvum* toxin interacting with bovine erythrocytes and causing haemolysis.

6.4 Platelets or Thrombocytes

Platelets are essential for the normal blood coagulation process. Their function is accomplished by adherence to each other (aggregation) and to the vessel where bleeding is taking place. Platelet function has previously been found to be dependent on cellular metabolism; therefore measurement of heat production rate (overall cell metabolism) offers the possibility to reveal metabolic derangements capable of causing functional abnormalities. Their concentration range in normal blood is 140–440 $\times$ 10^9 cells dm^{-3}.

6.4.1 Methodology

6.4.1.1 Preparation Procedures

Venous blood is collected in tubes containing an anticoagulant, citrate or heparin. Platelets are separated from other blood cells by repeated centrifugations. Significantly lower platelet heat production rate was found when heparin was used as compared to citrate (Monti and Wadsö 1977).

6.4.1.2 Storage, Temperature, Variations of pH and Cell Concentration

After 4–6 h storage at room temperature, the average decrease of heat production rate was 2% (Monti and Wadsö 1977). The heat production rate increases linearly with increasing temperature. The temperature coefficient was found to be Q_{10} = 2·0 at 37 °C for the temperature interval 32–42 °C. A similar effect was found by increasing pH, 0·02% increase of heat production rate per 0·01 pH unit at pH 7·4 (Monti and Wadsö 1977). Whereas in previous studies no significant concentration dependence was noted, in more recent observations it was found that the heat production rate decreases linearly with increasing cell concentration over the range 100–350 $\times$ 10^9 cells dm^{-3}, 0·54% per 10 $\times$ 10^9 cells dm^{-3} (Fagher and Monti 1984). Nanri and Minakami (1983) have studied pig platelets and found the heat production associated with glycolysis increased with pH, whereas the aerobic heat production decreased with increasing pH.

6.4.2 Basal Heat Production

Several investigators (Levin 1971; Bandmann *et al* 1975; Monti and Wadsö 1977; Fagher and Monti 1984) have found values about 60–70 fW cell^{-1} for platelets in the resting state (Table 6.3). The small variations in the results obtained by different authors are presumably related to different experimental conditions.

Table 6.3 Heat production rate in platelets from normal human subjects

Calorimeter	*Heat production rate* /fW cell^{-1} $\pm$ SD	*Source*
flow	68 ± 16	Levin (1971)
static ampoule	59 ± 8	Bandmann *et al* (1975)
static ampoule	61 ± 10	Monti and Wadsö (1977)
static ampoule	59 ± 4	Fagher and Monti (1984)

6.4.3 Stimulated Platelets

Fäldt *et al* (1982) found there was no increase of heat production rate when platelets were mixed with well defined immune complexes *in vitro*. The negative result is surprising, since it is known that platelets are stimulated metabolically when exposed to immune complexes.

Ross *et al* (1973) have studied platelet aggregation and found significant heat production when human blood platelets were exposed to various aggregating agents.

6.4.4 Clinical Studies

6.4.4.1 Thyroid Diseases

Platelet heat production rate was measured in hypothyroid patients by Valdemarsson *et al* (1985). Significantly decreased values were found before treatment in comparison with a group of healthy subjects (Table 6.4). Normalization of heat production rate was noted after treatment. These results show that decreased hormone production by the thyroid gland is accompanied by lowered thermogenesis in the platelets. This is in agreement with previous studies by Monti and Wadsö (1976a), where increased erythrocyte

Table 6.4 Heat production rate in platelets

Heat production rate/fW cell^{-1} $\pm$ SD			
Hypothyroidism		*Healthy subjects*	*Source*
Before treatment	*After treatment*		
51 ± 1	57 ± 2	59 ± 1	Valdemarsson *et al* (1985)
Patients with heart valve disease		*Healthy subjects*	*Source*
Before operation	*After operation*		
—	49 ± 9	65 ± 11	Monti and Fagher (1980)
63 ± 5	58 ± 5	59 ± 4	Fagher and Monti (1984)

thermogenesis was observed with increased thyroid function. In earlier investigations by Levin (1971) using a different calorimetric technique (flow), normal platelet heat production rate was observed in hypothyroid patients.

6.4.4.2 *Heart Valve Diseases*

Thromboembolism is a frequent complication after operative insertion of prosthetic cardiac valves. Although the cause is obscure, platelets have been suspected to play a significant role. In order to obtain information about possible cellular metabolic derangements, the heat production rate was measured in platelets of patients with prosthetic heart valves (Monti and Fagher 1980). It was found to be abnormally low (Table 6.4), as compared to the corresponding value in a control group. In a further investigation (Fagher and Monti 1984), a prospective study was done where a larger group of patients was randomly selected and heat production rates in their platelets before and after heart valve operation were determined (Table 6.4). Both studies suggest that insertion of prosthetic heart valves affects platelet metabolism. There is, however, an apparent discrepancy in the results obtained in the two trials. In the first study postoperative values are lower than normal, whereas in the second they are not. A possible explanation might be that the incidence of thromboembolism after the prosthesis operation was significantly higher (67%) in the first group (8 out of 12 patients) than in the second group, 21% (5 out of 24 patients). Young cells are metabolically more active and with higher aggregating capacity than older cells and therefore more apt to be consumed during thromboembolic episodes. Consequently, it can be presumed that a higher incidence of thromboembolism leads to the presence of an older cell population, reflected in the decreased metabolic activity. In fact, in both trials low rates of heat production were noted in samples from patients with thromboembolism.

6.4.4.3 *Malignant Hyperpyrexia*

In an attempt to develop a method for diagnosing susceptibility to malignant hyperpyrexia, the heat production rate was measured in platelets exposed to halothane and caffeine (Ranklev *et al* 1985). The results were negative. No differences could be found between susceptible and nonsusceptible individuals. However, in all cases the platelet heat production rate was lower in cell suspensions containing caffeine than without, thus showing a previously unknown effect of caffeine on platelet metabolism (Monti *et al* 1986b).

6.5 Lymphocytes

Lymphocytes represent one type of leukocytes or white blood cells. Their concentration range in normal blood is $1{\cdot}0$–$4{\cdot}8 \times 10^9$ cells dm^{-3}. They are a very important part of the immune system, whose function is the production of antibodies when challenged by antigenic materials.

6.5.1 *Methodology*

6.5.1.1 *Preparation Procedures*

Venous blood is collected in tubes containing an anticoagulant (heparin or citrate), or defibrinated by shaking the collected blood with glass beads. Lymphocytes are thereafter separated from other blood cells by density gradient centrifugation in Ficoll-Paque medium by the method of Böyum (1968). Phagocytizing cells (granulocytes and monocytes) still present are removed by incubating the cell suspensions in a phosphate buffer solution containing iron filings and removing the phagocytic cells with a magnet. A higher heat production rate was found in lymphocyte suspensions obtained from defibrinated blood than that in suspensions obtained from heparinized blood. Lymphocytes from heparinized blood also gave more reliable (steady state) calorimetric curves (Ikomi-Kumm *et al* 1984).

6.5.1.2 *Suspension Medium*

No significant difference was found when simultaneous calorimetric measurements were made on lymphocytes suspended in plasma or RPMI buffer solution containing sodium bicarbonate (Ikomi-Kumm *et al* 1984).

6.5.1.3 *Variation of pH and Temperature*

The heat production rate increases linearly with pH in the range 7·20–7·50. A similar variation with temperature was found in the range 25–37 °C, but not between 37 and 42 °C (Ikomi-Kumm *et al* 1984).

6.5.1.4 *Effect of Storage, Degree of Purity of Lymphocytes Preparation and Cell Concentration*

A decrease of heat production rate of about 5% per h was noted when cells were stored during 4 h at 25 °C. Unchanged values were found during 4 h storage of lymphocytes in a balanced salt solution or 0·9 g dm^{-3} NaCl without glucose (Ikomi-Kumm *et al* 1984). An increased heat production rate per lymphocyte was observed in the presence of phagocytic cells in the suspension (Ikomi-Kumm *et al* 1984). In the range 2–6 × 10^9 cells dm^{-3} the heat production rate per lymphocyte was constant, whereas in the lower concentration range 0·5–4·0 × 10^9 cells dm^{-3} slightly higher values per cell were found (Ikomi-Kumm *et al* 1984), thus suggesting a certain degree of 'crowding effect'. Lymphocytes have been previously reported to show a density dependent inhibition of O_2 consumption (Hedeskov and Esmann 1966).

6.5.2 *Basal Heat Production*

Variations of the heat production rate of human lymphocytes recorded by different investigators are presumably dependent on variable experimental conditions. Most studies, however, show values about 2 pW per lymphocyte

Table 6.5 Heat production rate in lymphocytes from normal human subjects

Calorimeter	*Suspension medium*	*Anticoagulant*	*Heat production rate*/pW cell^{-1} ±SD)	*Source*
static ampoule	plasma	defibrination	2·2 ± 1·4	Bandmann *et al* (1975)
flow	serum + medium 199	heparin	3–10[a]	Górski and Levin (1977)
static ampoule	plasma	defibrination	2·7 ± 0·5	Ikomi-Kumm *et al* (1984)
static ampoule	plasma	heparin	1·8 ± 0·3	Ikomi-Kumm *et al* (1984)

[a]Range: mean value and SD are not available.

(Table 6.5), while Krakauer and Krakauer (1976) have reported values of 8 pW per cell for horse lymphocytes.

6.5.3 Stimulated Lymphocytes

Although most studies on lymphocytes have been performed with cells in the resting state, there are few observations where heat production rate was measured when cells were stimulated. Krakauer and Krakauer (1976) have recorded the heat production rate in horse lymphocytes during stimulation with mitogenic agents. Increased heat production was observed after 2 days of incubation, and the thermogenic effect was found to precede DNA synthesis. Górski and Levin (1977) have reported similar results. Fäldt *et al* (1982), using batch calorimeters, have measured heat production rates in lymphocytes exposed to immune complexes and recorded a slight increase of heat production during the process.

6.5.4 Clinical Studies

6.5.4.1 Non-Hodgkin Lymphoma (NHL)

This disease is a malignant tumour of lymphoid cell proliferation. There are features in common with lymphocytic leukaemia. To NHL belong various groups of patients with different degrees of malignancy, difficult to distinguish clinically from each other. In order to treat optimally patients belonging to different groups, it is necessary to make a correct evaluation of their degree of malignancy before the treatment, more or less intensive, is started. At present there are no good ways to classify the different groups satisfactorily. The most accepted classifications are based on morphological features of the malignant cells. Microcalorimetry was applied by Monti *et al* (1981) to study the possible relationship between metabolism of malignant cells and clinical development

Table 6.6 Heat production rate in blood lymphocytes and tumour cells (lymphoid cells) from 21 patients with non-Hodgkin lymphoma. Cells were separated from heparinized blood and suspended in autochthonous plasma. Measurements were made in a static ampoule calorimeter

	Heat production rate/pW cell^{-1} ± SD	
Cell type	*Patients with remission*[a]	*Patients without remission*
blood lymphocytes	2·4 ± 0·5	3·7 ± 1·1
tumour cells	3·1 ± 0·8	5·5 ± 3·3

[a]Remission = responding to treatment.

of the disease. Heat production was measured in blood lymphocytes and tumour cells (lymphoid cells) in 21 NHL patients. Higher *p* values in both types of cells were found in the group of patients who did not respond to treatment (Table 6.6), as compared to the corresponding values in the group of patients improving during treatment. Of ten patients in the first group, seven had expired within 6 months. More recent studies (Monti *et al* 1985) in a group of 36 NHL patients with follow-up after about 5 years support the results of the previous investigation. These results suggest that in NHL, high cell metabolic activity is associated with a more malignant disease, and calorimetry might be of help in identifying patients in need of more intensive therapy.

6.5.4.2 *Chronic Lymphocytic Leukaemia (CLL)*

This is the most common type of leukaemia. Often hypermetabolism is present and causes various symptoms like loss of weight, rising temperature, sweating. The mechanism by which the hypermetabolic state develops is incompletely understood. Brandt *et al* (1979) have studied the energy balance in CLL patients by measurement of heat production in blood lymphocytes. The heat production rate per cell was not found to be increased, as expected. On the contrary, it was lower than normal. However, due to the increased number of lymphocytes, the estimated heat production of the total lymphocyte population was considerably higher than normal. The conclusion of this study was that the increase of heat production of the total lymphocyte population was of such magnitude that it may explain the symptoms of hypermetabolism present in this disease.

6.6 Granulocytes or Polymorphonuclear Neutrophils (PMNs)

Granulocytes represent one of the most numerous and important types of leukocytes. Their concentration range in normal blood is about 2–7 × 10^9 cells dm^{-3}. These cells have the very important function of ingesting (phagocytosis) and degrading foreign substances. Thus microorganisms are killed in infectious diseases.

6.6.1 Methodology

6.6.1.1 Preparation Procedures

Venous blood is collected in tubes containing heparin or citrate, or defibrinated by shaking the collected blood with glass beads. Blood is mixed with dextran-NaCl solution and the leukocyte-rich plasma is obtained. Granulocytes are thereafter separated from other blood cells by density gradient centrifugation in Ficoll-Paque medium by the method of Böyum (1968).

6.6.1.2 Suspension Medium

Bandmann *et al* (1975) have found that when granulocytes were suspended in plasma, the power decreased with time, whereas with cell suspensions in phosphate buffer solution the power remained constant.

6.6.1.3 Influence of Cell Concentration, Storage and pH Variation

No extensive methodological studies have been made for granulocytes as compared to other blood cells. Some results have, however, been reported in an earlier review (Monti and Wadsö 1979). The heat production rate per cell was found to decrease with increase in cell concentration, the 'crowding effect' (cf. the similar phenomenon for lymphocytes and platelets). Storage for 3 h at 4 °C or 24 °C did not affect heat production rate when granulocytes were resuspended in phosphate buffer solution. On the other hand, when cells were suspended in plasma the rate of production decreased by about 30% after storage at 24 °C for 3 h. A positive linear relationship was observed between heat production rate per cell and pH in the cell suspension.

6.6.2 Basal Heat Production

Bandmann *et al* (1985), using static ampoule calorimeters, have reported values of 1·3 pW per cell for granulocytes suspended in phosphate buffer solution and 3·5 pW per cell for plasma suspensions. Eftimiadi and Rialdi (1982) have obtained, using a flow calorimeter, values of 2·5 pW per cell for granulocytes suspended in Gey's saline solution.

6.6.3 Stimulated Granulocytes

Fäldt *et al* (1982) using batch calorimeters have measured heat production rates in granulocytes exposed to defined immune complexes formed *in vitro*. Large amounts of heat, corresponding to about 4–5 times the basal heat production, were recorded during activation. Furthermore, it was found that Fc receptors for IgG and C3b receptors were involved in the process. The results show that a new method, based on calorimetry, could be used for detection of immune complexes in blood. Eftimiadi and Rialdi (1982), using a flow calorimeter, have obtained similar results by measurement of heat production in granulocytes exposed to phorbol-12-myristate-13-acetate (PMA). These investigators have correlated heat production with O_2 consumption, ATP concentration, lactic acid production and CO_2 production from glucose oxidation. Their results show that most of the heat produced during activation depends on the aerobic pathway. Activation of granulocytes in the presence of immune complexes or PMA stimulates the so-called 'metabolic burst' that accompanies the important granulocyte function of phagocytosis.

6.6.4 Clinical Studies

The basal heat production rate in granulocytes from patients with acute myelogenous leukaemia was measured by Fäldt *et al* (1984) using static ampoule and perfusion calorimeters. Increased values of the rate of heat production were found in the acute stage of the disease and during the first 6 months, whereas after 6 months the values returned to normal. In addition, granulocytes from healthy subjects were found to have increased heat production rates when they were suspended in serum from leukaemic patients, thus suggesting that the increased metabolism in leukaemic granulocytes is serum dependent. It is not clear at present what factor in serum is responsible for granulocyte stimulation, but the possibility cannot be ruled out that immune complexes, often present in the blood of leukaemic patients, play a role in the process. Ankerst *et al* (1984) have found, using static ampoule calorimeters, that the responsiveness of granulocytes from patients with acute leukaemia to immune complex stimulation was decreased in the acute stage of the disease and during the first 6 months, whereas after 6 months the values returned to normal. The results seem to indicate that in patients with leukaemia the capacity of granulocytes to phagocytize foreign substances is impaired. In fact, it is known that infections are common in these patients, mostly related to decreased number of granulocytes but even defects in granulocyte function have been reported.

Eftimiadi and Rialdi (1984) have studied by flow calorimetry the heat of activation of granulocytes during phagocytosis of various strains of staphylococcus, and found great variations of granulocyte reponse to different staphyloccoci.

6.7 Leukocyte Mixtures

6.7.1 Basal Heat Production

Using a flow calorimeter and heparinized blood Levin (1971) has measured the heat production in plasma suspensions of platelet-leukocyte mixtures. By subtracting the calculated value for platelets the author calculated the value for leukocytes to be about 6 pW per cell. Using defibrinized blood and serum leukocyte suspensions Levin (1973c) has found a value of 2·2 pW per cell. This latter value is in fairly good agreement with the values obtained by Bandmann *et al* (1975) for granulocytes and lymphocytes.

6.7.2 Stimulated Leukocytes

Levin (1973c), using flow calorimetry, has also measured the increase in heat production during stimulation of phagocytosis by the presence of dead bacteria in the leukocyte suspension. A large increase of heat production was found, presumably due to the metabolic activation of granulocytes during phagocytosis. Similar results were obtained by Monti *et al* (1980) when phagocytosis was stimulated in whole blood by immune complexes.

6.7.3 Clinical Studies

6.7.3.1 Thyroid Diseases

Leukocyte heat production rate, measured with flow calorimetry, has been found by Levin (1971) to be increased in hyperthyroidism and decreased in hypothyroidism. Those results are in agreement with those of Monti and Wadsö (1976a) for erythrocytes from hyperthyroid patients and by Valdemarsson *et al* (1985) for platelets from hypothyroid patients. It is well known that thyroid hormones stimulate cell calorigenesis. However, the mechanism by which such an effect occurs is not completely understood. Usually, the effect of thyroid hormones on cell energy turnover is evaluated by measurement of oxygen consumption. It should not be forgotten, however, that anaerobic metabolism is probably as important and ought to be taken into account. In fact, for some cells, for example erythrocytes, most of the substrate is metabolized anaerobically. Measurement of heat production is therefore advantageous as it reflects the energy expenditure from the overall cell metabolism.

6.7.3.2 Immunological Diseases

Levin and Thomasson (1974) have found, using flow microcalorimetry, increased values in heat of activation, when leukocytes are stimulated by an homogenate prepared from their own leukocytes in patients with Systemic Lupus Erythematosus. According to the authors the method could be used to assess the degree of activity in immunological diseases.

6.8 Whole Blood

Heat production has been measured in whole blood and found to be about $60\,\mu W\,cm^{-3}$ at 37 °C and pH 7·4 (Bandmann *et al* 1975). Comparative measurements were made at the same time on the different cell populations. About 50% of the heat produced was due to the erythrocyte metabolism, whereas lymphocytes and granulocytes accounted for 20% each, and platelets produced about 10% of the total heat production. The sum of the values for the different cell fractions was in good agreement with the experimental value for the whole blood. Using batch microcalorimeters, the heat production of human blood was measured in the presence of antigens and their specific rabbit antisera; it was greatly increased, indicating a metabolic activation of blood cells (Monti *et al* 1980). The results suggest that calorimetry may be a useful technique for detection of antigen-antibody complexes.

References

Ankerst J, Fäldt R and Monti M (1984) *Leukemia Res* **8**, 997–1002
Bandmann U, Monti M and Wadsö I (1975) *Scand J Clin Lab Invest* **35**, 121–7
Binford J S Jr, Martin D F and Padilla G M (1973) *Biochim Biophys Acta* **291**, 154–64
Boyo A and Ikomi-Kumm J (1972) *Lancet* **1**, 1215–6
Brandt L, Ikomi-Kumm J, Monti M and Wadsö I (1979) *Scand J Haematol* **22**, 141–4
Brewer G J (1970) *Red Cell Metabolism and Function*. New York and London, Plenum Press, p 103
Böyum A (1968) *Scand J Clin Lab Invest* **21** (suppl), 97
Cerny L C, Cook F B and Stasiw D M (1974) *Biorheology* **11**, 235–9
de Verdier C-H (1981) *Acta Biol Med Ger* **40**, 669–702
Eftimiadi C and Rialdi G (1982) *Cell Biophys* **4**, 231–44
Eftimiadi C and Rialdi G (1984) *J Infect Dis* **150**, 366–71
Fagher B and Monti M (1984) *Eur Heart J* **5**, 55–61
Fäldt R, Ankerst J, Monti M and Wadsö I (1982) *Immunology* **46**, 189–98
Fäldt R, Ankerst J and Monti M (1984) *Br J Haematol* **58**, 671–8
Górski A and Levin K (1977) *Science Tools* **24**, 27–9
Grarup J, Hasselager E and Hjelm M (1977) *Scand J Clin Lab Invest* **37**, 85–8
Hedeskov C J and Eskmann V (1966) *Blood* **28**, 163–74
Ikomi-Kumm J (1977) *IRCS Med Sci* **5**, 320
Ikomi-Kumm J and Boyo A (1977) *IRCS Med Sci* **5**, 576
Ikomi-Kumm J (1984) Thesis, Lund, Sweden
Ikomi-Kumm J, Monti M and Wadsö I (1984) *Scand J Clin Lab Invest* **44**, 745–52
Jones M N and Zala C (1974) *Proc R Soc Lond* [*Biol*] **185**, 73–82
Krakauer T and Krakauer N (1976) *Cell Immunol* **26**, 242–53
Levin K (1971) *Clin Chim Acta* **32**, 87–94
Levin K and Boyo A (1971) *Scand J Clin Lab Invest Suppl* **118**, 55
Levin K (1973a) Thesis, Västerås, Sweden
Levin K (1973b) *Scand J Clin Lab Invest* **32**, 55–65
Levin K (1973c) *Scand J Clin Lab Invest* **32**, 67–73
Levin K, Fürst P, Harris R and Hultman E (1974) *Scand J Clin Lab Invest* **34**, 141–8
Levin K and Thomasson B (1974) *Acta Med Scand* **195**, 191–200
Levin K (1980) In: Beezer A E (ed) *Biological Microcalorimetry*. London, Academic Press, pp 131–43
Minakami S and de Verdier C-H (1976) *Eur J Biochem* **65**, 451–60
Monti M (1977a) *Scand J Haematol* **18**, 154–62
Monti M (1977b) *Scand J Haematol* **19**, 313–8
Monti M and Wadsö I (1973) *Scand J Clin Lab Invest* **32**, 47–54

Monti M and Wadsö I (1976a) *Acta Med Scand* **200**, 301–8

Monti M and Wadsö I (1976b) *Scand J Clin Lab Invest* **36**, 431–6

Monti M and Wadsö I (1976c) *Scand J Clin Lab Invest* **36**, 565–72

Monti M and Wadsö I (1976d) *Scand J Clin Lab Invest* **36**, 573–80

Monti M and Wadsö I (1977) *Scand J Haematol* **19**, 111–5

Monti M and Wadsö I (1979) In: Jones M N (ed) *Biochemical Thermodynamics*. Amsterdam, Elsevier, pp 256–80

Monti M and Fagher B (1980) *Clin Chim Acta* **102**, 83–9

Monti M, Fäldt R, Ankerst J and Wadsö I (1980) *J Immunol Methods* **37**, 29–37

Monti M, Brandt L, Ikomi-Kumm J, Olsson H and Wadsö I (1981) *Scand J Haematol* **27**, 305–10

Monti M and Ikomi-Kumm (1985) *Metabolism* **34**, 183–7

Monti M, Brandt L, Ikomi-Kumm J and Olsson H (1986a) *Scand J Haematol* **36**, 353–7

Monti M, Edvinsson L, Ranklev E and Fletcher R (1986b) *Acta Med Scand* **220**, 185–8

Monti M, Hedner P, Ikomi-Kumm J and Valdemarsson S (1986c) *Metabolism* (in press)

Nakao M, Makayama T and Kankura T (1973) *Nature New Biology* **264**, 94

Nanri H and Minakami S (1983) *Biomed Biochim Acta* **4**, 315–21

Ranklev E, Monti M and Fletcher R (1985) *Br J Anaesth* **57**, 991–3

Ross P D, Fletcher A P and Jamieson G A (1973) *Biochim Biophys Acta* **313**, 106–18

Suurkuusk J and Wadsö I (1982) *Chemica Scripta* **20**, 155–63

Valdemarsson S, Fagher B, Hedner P, Monti M and Nilsson-Ehle P (1985) *Acta Endocrinol* **108**, 361–6

Wadsö I (1974) *Science Tools* **21**, 18–21

Chapter 7

Heat Dissipation and Metabolism in Isolated Animal Cells and Whole Tissues/Organs

R. B. Kemp

7.1 Introduction

When last I reviewed this topic (Kemp 1980), hard numbers for heat production by isolated cells rather than whole tissues and organs were only just emerging because of technical problems. Highly sensitive calorimeters had been designed with chemistry not biology in mind, naturally, and many biologists who fed their material into calorimeters 'because they were there' only achieved a frightful mess (see Kemp 1972). Thus, aspirants in the 1970s spent a good deal of their time trying to make the instruments and tissues or cells compatible and divorce was often the only result. Leaving aside plain ignorance, it is a simple fact that animal cells are not as easy to work with as most microbes; that is, unless one takes a microbiologist's standpoint! Animal cells are more fragile and generally more sticky. They do not usually grow and divide in suspension. The maximum cell density in culture is low in relative terms and, generally, the nutritional requirements to achieve this density are extremely complex. It is not surprising, then, that progress has been slower than hoped and considerations of instrumental modifications uppermost (Kemp 1980). Now is an appropriate moment, however, to consider the positive returns for all the early groundwork, in terms of its contributions to our knowledge of tissues and cells (other than blood—p. 131). But woe betide the newcomer who neglects instrument and, particularly, vessel design (p. 49).

Practically all of the studies involving heat measurements of tissues have been short term and this is partly true for isolated cells; thus avoiding the complications of cell growth and division in terms of culture requirements, including sterility. The exceptions will be fully highlighted and comprehensive considerations will also be given to the two other major, but not by compulsion exclusive, applications of calorimetry, namely heat profiles ('thermograms') and the effect of exogenous materials on them, and the relation of power-time curves to cellular metabolic pathways. It is worthwhile reiterating the point that the above divisions are made for the catalogue and not because the essential elements of each type of study cannot be combined—indeed, perhaps the contrary ought to be a resolve!

7.2 Long Term Studies of Cellular Heat Production

This type of investigation has given researchers more headaches than test data because cells in culture are most particular about their environment. In 1975, Kemp placed monolayers of chick embryo fibroblasts (primary culture) established on glass beads in the microcolumn of an LKB flow sorption calorimeter. The cells were perfused with culture medium, including 10% (v/v) calf serum. There was an overall linear increase in power for the first 16 h, the quantity doubling in 12·5 h—a time which closely corresponded to the generation cycle. From 16 h onwards, there was a gradual decrease in heat production and little cell growth and division. Thus, heat evolution (2·02 pW $cell^{-1}$) only compared favourably with that of freshly dissociated cells (2·3 pW $cell^{-1}$) over the first period. Despite several modifications in different experiments, cultures were never confluent and so one must assume that conditions in the sorption vessel were not adequate.

At that time, it appeared difficult to conduct this type of measurement in an LKB calorimeter and, indeed, this view was reinforced by the findings of Cerretti *et al* (1977). They used an LKB batch instrument equipped with glass vessels which, once charged with cells, were rotated every 30 or 60 min over a 24 h period. Apart from estimating that suspension-adapted KB and HeLa-S3 cells both evolve heat at 25 pW $cell^{-1}$ with the population doubling in 24 h, the authors reported large thermal deflections with each programmed rotation of the block. Mixing KB cells initially caused a rapid endothermal disturbance, the extent of which was dependent on the interval between rotations. This signal was immediately followed by an exothermal deflection with a duration as long as 30–60 min. The magnitude appeared to be dependent on the number of cells *not* attached to the glass, a supposition confirmed by experiments using HeLa cells in which, if the majority were adherent to the 'substrate', only the endothermic deflection occurred on rotation. This maelstrom of deflections was explained in terms of 'crowding effects' and, interesting as this is, there is nothing more frustrating to a 'deltaman' than a phenomenon!

It is suspected that a similar problem was encountered by Ljungholm *et al* (1978) who measured the heat produced by HeLa-S3 cells using an ampoule drop calorimeter (Spink and Wadsö 1976), similar to that manufactured by LKB. Cells in growth medium were incubated for 16–18 h in a sterilized stainless steel ampoule to allow attachment. Just prior to inserting the vessel in the calorimeter, the medium was changed to one containing only 0·25% (v/v) calf serum, supposedly to force the cells into the same growth phase—submission by starvation! Although this procedure did result in a greater degree of reproducibility than found in preliminary experiments using a full culture medium (10% calf serum), the power varied between 20 and 40 pW cell^{-1}; a variation which is clearly undesirable and could only be explained by the authors as 'depending upon the state of the cell'.

Nicolic and Neskovic (1976) took a different approach to the problem by placing a standard culture dish in the instrument rather than using commercially available vessels designed for other purposes. This was possible because the design of the Calvet calorimeter (Setaram) allowed easy insertion and removal of the dish. L929 mouse fibroblasts in a standard culture medium with calf serum, were seeded in a hermetically sealed glass vessel and formed a monolayer, the cells having a generation time of approximately 12 h. Nicolic and Neskovic (1976) did not give a value for power per unit cell but, taking the dangerous course of estimating from their power-time plot, the figure appeared to be 69 pW cell^{-1} for those growing from a small inoculum ($2{\cdot}5 \times 10^5$ cells cm^{-3}) after 18 h (Kemp 1980). Unfortunately, these investigators seem not to have continued work in this field and it was left to Schaarschmidt and others to take up the cudgels.

Measurement of heat dissipation by human skin biopsies (Hellgren and Larsson 1977) and epidermal slices (Anders *et al* 1977, 1979) had met with reasonable success despite technical problems (see Schaarschmidt and Reichert 1985) but, in 1980, Pätel *et al* succeeded in maintaining long term cultures of isolated human skin fibroblasts in a Bioflux differential microcalorimeter (Thermanalyse) which had two 18 cm^3 stainless steel removable vessels. They seeded human foreskin fibroblasts (Flow finite line 7000) onto plastic foil (70 × 50 mm) which had been removed from a Sterilin bulk cell culture vessel and sterilized with absolute ethanol. A period for cell attachment was allowed before the foil was inserted into the sterile calorimetric vessel, together with growth medium containing 15% (v/v) calf serum. A culture of 3×10^6 cells gave a power-time curve with an initial maximum over the first 3 h, followed by slightly increasing heat production for 10–20 h. Power then decreased to a lower value of 30–40 pW cell^{-1}, later revised to 40 ± 10 pW cell^{-1} (Schaarschmidt and Reichert 1981), which was maintained for several days. The first peak was ascribed to spreading and growth of the cells, a supposition borne out by the fact that it was not found if the time between seeding and the first calorimetric response was 4–5 h (Schaarschmidt and Reichert 1981). The increasing heat evolution that followed was ascribed to cell multiplication because it corresponded to an increase in cell count, as confirmed using parallel cultures outside the calorimeter. Under the particular conditions of this study, the culture became confluent after 2–3 divisions and heat production declined to a constant value, supposedly because metabolism was 'restricted to processes of maintenance'.

This research group has recently extended its successful studies to another skin component, the keratinocyte. This cell type does not develop a similar terminal pattern of differentiation *in vitro* on glass or plastic surfaces to that found *in vivo* (Schaarschmidt and Reichert 1986). However, advantage was taken of a recent development, the Petriperm culture dish with a flexible hydrophilic membrane, which allows nutrient and oxygen exchange (Reichert and Schaarschmidt 1986). Trypsin-dissociated keratinocytes from human breast skin were grown to confluency in dishes coated with acid-soluble collagen. The membrane was cut out and placed in the calorimeter in the same manner as described for skin fibroblasts. Of course, their proliferative capability was low owing to contact inhibition of growth and division and, probably for this reason, the power-time curve was less complex than that for fibroblasts. There were small peaks in the first hour which the authors attributed to the introduction of fresh culture medium but, within 5 h, the curve declined to a constant level representing 83 ± 12 pW cell^{-1}, which was maintained for at least 18 h. Power was double that of skin fibroblasts and, according to Reichert and Schaarschmidt (1986), represented 'an equilibrated energy metabolism'. They also cultured an SV-40 transformed human foreskin keratinocyte line, SV-K14, in a similar fashion, although without collagen. The power-time curve had the same form but heat evolution was higher, 134 ± 35 pW cell^{-1}, probably owing to the fact that transformed cells do not exhibit contact inhibition. Under confluent conditions, 'surplus' cells, like lemmings, detach and die!

Now, pretty as some long term power-time curves (thermograms) for cultured cells are, analysis can only be speculative because of the obviously complex energetics of growth and division on substrates in non-defined media, that is those containing calf serum. To give one example, Pätel *et al* (1980) thought that the first peak in their curve was due to attachment and spreading, but Pätel (1981) noticed that it occurred even if the foil-culture was placed in the calorimeter long after completion of that phase. Peak height was above 300 μW, double that reported a year later (Pätel *et al* 1982) although, in both papers, the quoted power was $51{\cdot}5 \pm 5$ pW cell^{-1} (higher than reported by Schaarschmidt and Reichert 1981) for active, but not dividing cells. By this time, a tenter frame was being used to support the foil, but one is left on tenter-hooks! Pätel (1981) thought that the first phase of the curve was attributable to a change from aerobic to anaerobic conditions, an assumption which could be tested simply by analysis of intermediary metabolism. Nevertheless, if the assumption is correct, then conditions in the calorimeter are not ideal for cultured cells; although it is only fair to point out that, earlier, Schaarschmidt and Reichert (1981) did conduct tests to show that metabolism in such a system was not limited by oxygen.

Recent technological advances in Wadsö's laboratory (p. 34) give hope of a solution to the technical problems of calorimetric conditions for cell culture. Following design of a multichannel calorimetry system (Suurkuusk and Wadsö 1982), now commercially available from LKB, attention turned to the notion of a modular vessel with stirrer shaft, gas and/or liquid flow and different types of blades and sample holders. In the report of a successful vessel, Görman Nordmark *et al* (1984) detail technical trials with cells. Human T-lymphoma cells (CCRF) in a stirred suspension medium without calf serum (plus gas phase) produced heat at 6·1 pW cell^{-1}. In the second experiment, anchorage-dependent H1477 human melanoma cells were grown to confluency on a 'Schaarschmidt-style' polystyrene film (30 × 30 mm). This was made into a cylindrical shape with cells facing inward and inserted into a 3 cm^3 sample cup. The vessel was charged with culture medium (no calf serum), which also served to perfuse the vessel during the period of measurement (10 cm^3 h^{-1}, reversed flow). The cells adherent to the film (0·9 × 10^6) gave a power of 73 μW, about 80 pW cell^{-1}. Although these experiments on cell lines were essentially short term trials, this new vessel has an obvious importance for the thermal analysis of cellular maintenance, growth and division.

Studies of the effects of external agents on the heat profile of growing cells have not progressed too far, for the above-stated reasons, as well as for the fact that, in many cases, there is only a requirement for a short period of measurement, for instance metabolic inhibitors (see p. 156). The converse is true, however, for the interaction of microbes with cells in which there can be a train of events. Ljungholm *et al* (1978) studied the effect of adenovirus (type 5) infection on heat evolution by HeLa cells and found markedly different profiles, especially in the timing of maxima for power. Infected cells displayed a peak 4 h before that of controls (12 h). Production of viral capsomers was not thought to have started by that time but there was a rapid decline in heat evolution, indicating that cellular metabolism was affected considerably before maturation of virions. In this study, as with the following one, long term culture is an essential prerequisite.

Prokaryotes of the class Mollicutes—mycoplasma—are the ruin of many a cell culturist. Infection is often covert, so can lead to erroneous results. Flanagan and Kemp (1986) investigated the effect of *Mycoplasma bovirhinus* (Leach 1967), strain PG43, on LS-derived L929 mouse fibroblasts in suspension culture. The cell suspension (9 × 10^5 cells cm^{-3}) was pumped at 20 cm^3 h^{-1} from a bulk culture flask, through a modified flow vessel (Kemp's 'matchbox', 1980) in an LKB (old style) flow calorimeter. After establishing a plateau corresponding to 34 $\pm$ 3 pW cell^{-1}, the culture was infected with mycoplasma (5 × 10^6 cells cm^{-3}). There was a small but rapid (allowing for dead space) increase in power (17%), corresponding to the heat production by mycoplasma and similar to that found by Ljungholm *et al* (1976). Although mycoplasma have a rapid generation time (2 h), the power-time curve settled to a plateau just below the maximum power for 10 h. Then there was a gradual decline in heat produced by the infected culture until at 16 h power was 33% of maximum and cell count 18% of the original figure. *M. bovirhinus* binds to cells—this could be studied using a flow mixing vessel—and appears to cause cytolysis, probably by local effects on the plasma membrane of acid metabolites from glycolysis.

The effect of drugs on heat production by established cultures has great potential both in terms of mode of action and as a screen. Pätel *et al* (1980) first showed that anthralin, a drug used to treat psoriasis, caused a decrease in heat output of 30%. Then, in a series of papers Pätel (1981) and Pätel *et al* (1981, 1982) established the dose-response (heat) curves for different derivatives of anthralin and were able reliably to discriminate between them. They also discovered that the respiration (as measured by manometry) of skin fibroblasts is more sensitive to anthralin than heat production. If this finding was associated with other data on catabolic pathways in these cells, perhaps a clue to the mechanism of action by the drug would emerge—it was not understood in 1980 (Pätel *et al*)!

It may have been noticed that, throughout this section, attention has been drawn to the amount of animal serum in the culture medium of the cells under investigation. None of the cells in question has been adapted to live in non-sera media, so the reduction in or absence of serum must cause metabolic changes and stress, probably leading to qualitatively and quantitatively different power-time curves. Of course, serum is a nightmare 'witches brew' for both cell culturists—batch variations, source of infection, to mention a couple—as well as to thermal analysts, who may, and perhaps should, wish to relate calorimetric data to cellular metabolic pathways. Thus, it would be advisable to adapt established cell lines to non-serum media, although this can be an arduous task; heat output could be one of the parameters employed to monitor the adaptation process. In fact, this is precisely what is happening in our laboratory at the present time.

Although no specific mention has been made of it, all of the studies reviewed in this section have involved using a pot-pourri of antibiotics in the culture medium. It is well known that these substances, as well as minimizing the likelihood of infection, can exert significant effects on cellular metabolism and, therefore, heat (Gnaiger 1981). Researchers would be advised critically to assess the requirements for antibiotics and to evaluate their influence on metabolism, when interpreting data.

7.3 Short Term Thermograms and the Effects of Diverse Agents

This type of study attracted early prominence because it seemed that calorimetry would offer a novel and non-destructive means of investigating cellular interaction with a motley host of substances. Insight has varied from clear to opaque but, in all cases, there is promise.

Perhaps for no better reason than a vested interest, let us begin with fat cells. In 1980, Monti *et al* dissociated human subcutaneous adipose tissue into adipocytes using collagenase ($3{\cdot}3\,g\,dm^{-3}$) in Krebs-Ringer-bicarbonate buffer containing albumin ($30\,g\,dm^{-3}$), insulin ($0{\cdot}1\,U\,cm^{-3}$) and glucose ($11\,mmol\,dm^{-3}$). Cells were then suspended in the same buffer solution (pH 7·4) containing insulin and glucose and placed in the stainless steel cylinder of an ampoule microcalorimeter (Spink and Wadsö 1976; Wadsö 1976). Under these conditions and at 37 °C, heat evolution was 49 ± 15 (SD) $pW\,cell^{-1}$. Power increased with pH between 7·14 and 7·68 and with temperature between 32 and 42 °C ($Q_{10} = 1{\cdot}9$). The absence of glucose predictably affected heat production and insulin alone had no influence. Omission of both reduced power by 50%. To bear out a point made earlier autologous serum (10–15%) in the suspension medium caused a 20% increase in heat evolution. It is also relevant to signpost the roles of two types of agent, a metabolite (glucose) and a hormone (insulin), in total cellular power output.

Having established reasonable cell culture conditions, this group compared heat production by adipocytes from obese and normal weight individuals (Sörbris *et al* 1979). Cells from obese humans surprisingly gave a significantly lower power value ($26 \pm 12\,pW\,cell^{-1}$) than that of 'normals', already stated above. The observation that adipocytes from obese subjects were markedly larger than those of controls minimizes this difference. Thus, it was better emphasized in terms of $\mu W\,g^{-1}$ fat tissue—40 for obese against 133 for controls. The investigators stated that 'there is increasing evidence that obesity is associated with an increased metabolic efficiency'. Now, this is not a particularly new idea for, 30 years ago, a corpulent Latin master said, when criticizing a typically poor translation of Ovid, 'Kemp, you and I have only one characteristic in common—a highly efficient digestive system!' Little could he have prophesied that Sörbris *et al* (1979) would speculate that the efficiency resulted from restricted activity of 'futile' (substrate) cycles—as in the Classics!

In a more recent study, Sörbris *et al* (1982) persuaded their obese patients to lose weight—an average 17 kg (12%) in 5 weeks. Power output of adipocytes increased from 25 to $38 \pm 17\,pW\,cell^{-1}$ ($P < 0{\cdot}05$), which was still significantly lower ($P < 0{\cdot}01$) than the previously obtained figure for controls. The cells, like Alice, became smaller (by 10%) and appeared to be more sensitive to insulin. This research group has now resorted to the more drastic procedure of surgical therapy—horizontal gastroplasty (Sörbris *et al* 1986) The mean body weight of 15 patients was reduced by 30% to 97 kg, 6–8 months after surgery and adipocytes were smaller by 33%. Heat production ($66 \pm 9\,\mu W\,g^{-1}$ fat tissue) prior to surgery was higher than that of an earlier obese group (Sörbris *et al* 1979) but became $115 \pm 7\,\mu W\,g^{-1}$ fat, 6 months after operation. This was still less than the value for controls although, when expressed per cell, heat production did not differ from that found in lean subjects. Associated with weight reduction was a 50% decrease in plasma insulin levels and a 30% fall in fasting blood glucose concentration. The authors concluded that, although decreased heat evolution by adipocytes from the obese was a secondary phenomenon, there was an inverse correlation between cellular power and body weight—a point requiring elucidation by studies during the dynamic phase of weight gain. One wonders if they pay well!

Perhaps they do not, for Nilsson-Ehle and Nordin (1985) have tackled the problems by inducing experimental obesity in rats. Measurements with an LKB bioactivity monitor gave a value of 40 pW cell^{-1} for epididymal adipocytes under basal conditions. As with human adipocytes, insulin alone had no effect, 5·5 nmol dm^{-3} glucose raised power by 100% and the two together increased heat production by approximately 250%. Cells from rats induced to obesity by VMH lesion showed a 4- to 5-fold increase in heat evolution, which was related to greater incorporation of glucose into glycerides.

Turning from bad fat to good fat, it should be recalled that Nedergaard *et al* (1977) employed an LKB batch microcalorimeter to register heat production by isolated cells from hamster brown adipose tissue. The basal heat rate was 110 pW cell^{-1}, higher than that for white adipocytes, and 820 pW cell^{-1} for brown fat cells maximally stimulated by noradrenaline in the process of fat oxidation. It is interesting to note that Nilsson-Ehle and Nordin (1985) found that noradrenaline greatly stimulated heat production by white adipocytes, while Giradier *et al* (1976) reported that the same hormone caused an increase in the heat evolution of rat brown adipose tissue from the basal rate of 3·6 to 18·9 mW g^{-1}. Furthermore, it has been shown that cells dissociated from white fat originating in different regions of the rat varied up to threefold in heat production (Nilsson-Ehle and Nordin 1985). The scale of perirenal > retroperitoneal > epididymal > subcutaneous closely resembled the distribution of uncoupling protein, an indicator of brown adipose cells, in white fat deposits.

The Swedish researchers have recently turned their attention to thermogenesis by adipocytes and platelets (p. 136) from hypothyroid patients, using an ampoule calorimeter (Valdemarsson *et al* 1985). The heat produced by isolated fat cells from ten individuals with primary hypothyroidism was significantly lower (19 ± 2 pW cell^{-1}; $P < 0{\cdot}001$) than normal (49 ± 4 pW cell^{-1}). After therapy for 3 months on full *l*-thyroxine substitution, the mean value for power was 33 ± 3 pW cell^{-1}, significantly higher ($P < 0{\cdot}025$) than before treatment but still below the figure for controls ($P < 0{\cdot}01$). The increase in heat produced by the cells after treatment correlated well ($r = 0{\cdot}84$; $P < 0{\cdot}05$) with the gain in serum S-triiodothyronine (S-T_3) levels. The authors concluded that, in hypothyroidism, total metabolic activity as measured by calorimetry seems to be comparatively more reduced in adipocytes than in platelets.

Enough of fat and hormones and things! Hoffner *et al* (1985) utilized a cell line, L929 mouse fibroblasts, adapted to grow in suspension rather than

Krakauer 1976; Monti, p. 140). These cells suspended in a complex medium were stimulated for 3–5 days by either the specific antigen, dinitrophenylated bovine serum albumin, or the relatively non-specific mitogen, concanavalin A. Cycloheximide, (1 mmol dm^{-3}) not only prevented stimulation but reduced heat evolution to below that of the untreated controls. Of course, they had similar problems when using 10 $\mu g\,cm^{-3}$ cytochalasin B as an inhibitor of mitogenesis through its action on the cytoskeleton. Heat production was severely depressed, probably because of interference with transmembrane transport.

While these cautionary tales must be told, the positive aspect is that heat measurements do reveal the transitions of cellular states to different levels of activity, providing that other conventional biochemical assays are undertaken in parallel. Thus, there is the prospect of mechanistic studies on the action of drugs and toxic substances on target cells. Gnaiger (1981) has pointed the way in his calorimetric investigation into the effects of streptomycin and neomycin on aquatic animals. More recently, Thorén and Holma (1986) exposed rabbit alveolar macrophages in monolayer to different metal and non-metal oxide particles, and measured their heat production using an LKB bioactivity monitor. Over a 24 h period, particles of titanium oxide had no effect on heat output, and thus were considered non-toxic to the cells. On the other hand, manganese(IV)oxide was cytotoxic, there being a good correlation between reduced cell viability and less generated power. As a further twist to the particle story, the authors found that quartz induced an increase in heat production—a higher metabolic rate, they stated—reminiscent, probably, of the increased energy expenditure by granulocytes (neutrophils) during phagocytosis (Eftimiadi and Rialdi 1984, 1985a,b, 1986; Monti, p. 141).

It is not difficult to envisage the great potential of heat measurements in medicine, toxicology and pharmacology, as well as in basic sciences; but microcalorimetry must be associated with biochemical data in order for it to achieve its true worth.

7.4 Combination of Heat Measurements with Biochemical Assays

Although calorimetric measurements detect changes in enthalpy of all cellular reactions, biochemical data are required in order to obtain a physiological interpretation of these alterations. The most common and necessary association is with determinations of oxygen consumption—indirect calorimetry (Nedergaard *et al* 1977; Eftimiadi and Rialdi 1982; Hoffner *et al* 1985). The earliest calculations of the caloric equivalent of oxygen consumption were based on enthalpies of combustion estimated by Ivlev (1934) using a bomb calorimeter. His oxycalorific coefficient was quoted by Nedergaard *et al* (1977) when they showed that calculated and measured ratios of heat evolution to oxygen consumption in hamster brown adipocytes were consistent with expectation.

In more recent times, Gnaiger (1980, 1983a) has calculated what he termed the oxycaloric equivalent in dissipative metabolism, $\Delta_k H_{O_2}$, from chemical stoichiometries and the corresponding enthalpies of reaction. He defined 'dissipative' as the heat leaving an organism (Gnaiger 1983b) and, of course, this can be extended to isolated cells—hence, the involvement of direct calorimetry. The precise value of the theoretical oxycaloric equivalent for a given cell type in specific culture conditions is obviously dependent on the character and relative importance of operative metabolic pathways. As Hammerstedt and Lovrien (1983) stated, this requires a quantitative description of the way the cell partitions nutrients to form the many metabolic end products; in other words, classical assays of intermediary metabolism. In summary, the relationship between oxygen uptake measurements, $\dot{N}_{O_2}$, rates of heat dissipation, $\dot{Q}$, and *appropriately* derived oxycaloric equivalents, $\Delta_k H_{O_2}$, is:

$$\Delta_k \dot{Q} = \Delta_k H_{O_2} \times \dot{N}_{O_2}$$

where the suffix 'k' stands for the 'catabolic half cycle' (Gnaiger 1980); that is, dissipative conversion of substrates to CO_2, H_2O and nitrogenous end products without net storage of high energy intermediates.

Having written the above, it is now the point to confess that only three groups have conducted sustained 'combination' studies and one of these, that of Rialdi, is more properly discussed elsewhere (p. 143). Strangely enough, one of the two remaining teams has not routinely availed itself of oxygen uptake measurements. This is probably because the principal concern has been towards the importance of endogenous metabolism to total cellular metabolism under normal conditions and stress, as simply represented by the ATP cycle (Hammerstedt and Lovrien 1983). Requirements for ATP are met by blending the degradation of exogenous and endogenous substrates, and consequently it is important to know the rate of endogenous metabolism in making estimates of the total rate of ATP synthesis.

While, under normal physiological conditions, synthesis is balanced by ATP consumption, successful adaptation to the stress of anaerobic metabolism involves changes in adenylate concentrations and phosphagen levels (Gnaiger 1983b). These changes have been summarized as ergobolic reactions, which are coupled to catabolic or anabolic half cycle reactions. All this means in thermodynamic terms is that, during the transition from aerobic to anaerobic conditions, ergobolic heat changes (exothermic or endothermic) must be added to the catabolic heat changes (Gnaiger 1983b). So much for the simple ATP cycle!

Hammerstedt and Lovrien (1983) chose sperm as a model system in their studies of endogenous metabolism because biosynthesis, growth and alternate pathways for glucose metabolism (e.g. glycogen synthesis and pentose phosphate pathway) are insignificant. They described a heat conduction batch calorimeter of the distinctive Lovrien marque but left the details of their experimentation to another paper (Inskeep and Hammerstedt 1983). They incubated bovine sperm at 30 °C in a salt solution containing antibiotics and a non-metabolized buffer, 30 $mmol\,dm^{-3}$ endo-norbornene-5,6-*cis*-dicarboxylic acid, for 30 min. Using standard isotopic tracer methods after adding [^{14}C] glucose, it was shown that over 95% of the glucose carbon was recovered as lactate, acetate (including acetylcarnitine) and CO_2. The heat produced during these conversions was calculated using standard enthalpies of reaction and accounted for 90% [-412 ± 34 (SEM) $mJ\,h^{-1}$ (10^8 cells)$^{-1}$*] of the total measured heat production. The authors ascribed the remaining percentage heat [-43 ± 14 $mJ\,h^{-1}$ (10^8 cells)$^{-1}$] to metabolism of endogenous reserves. This assertion was supported by the finding that 4·9 $\mu mol\,dm^{-3}$ antimycin plus 9·0 $\mu mol\,dm^{-3}$ rotenone reduced endogenous metabolism to a very small level [-6 $mJ\,h^{-1}$ (10^8 sperm)$^{-1}$], which was consistent with data indicating that phospholipids are the endogenous substrate in sperm, the pathway involving β-oxidation.

Turning attention to high energy intermediates and thus ergobolic reactions, Inskeep and Hammerstedt (1983) found that, although total adenine nucleotide content [55 nmol (10^8 cells)$^{-1}$] was unchanged during the 30-min incubation, total energy charge increased from 0·62 to 0·82; but only in the absence of electron transport inhibitors. In terms of thermodynamic analysis, this difference in states could be troublesome. They have calculated, however, that the total heat produced by the redistribution of adenine nucleotides was $+0{\cdot}53$ mJ (10^8 cells)$^{-1}$,* a negligible value in the total. They went on to estimate that ATP produced in sperm by endogenous metabolism in the presence of glucose was approximately 10% of the total yield.

Following this clear demonstration of the value of calorimetry in assessing the relative importance of endogenous metabolism, Inskeep and Hammerstedt (1985) then investigated the role of the latter in total ATP synthesis by sperm at different temperatures (20, 25, 30 or 35 °C) and with various exogenous substrates (glucose, fructose, lactate or β-hydroxybutyrate)—the catabolic rate linked to ergobolic reactions. Sperm incubated at the different temperatures showed slight significant differences in heat production [-98 ± 17 $mJ\,h^{-1}$ (10^8 cells)$^{-1}$ at 25 °C and -144 ± 17 $mJ\,h^{-1}$ (10^8 cells)$^{-1}$ at 35 °C] but there was no change in total adenine nucleotide content or energy charge. There was, however, a gradual decrease in the rate of heat evolution over the 30-min period. In the presence of each of the substrates, heat dissipation increased with temperature [for instance, in $mJ\,h^{-1}$ (10^8 cells)$^{-1}$ and at 20 and 35 °C respectively, glucose incubation gave -200 and -623 and lactate -168 and -348]. In addition, heat production by sperm

*1 $mJ\,h^{-1}$ (10^8 cells)$^{-1} \equiv 2{\cdot}78 \times 10^{-15}$ $W\,cell^{-1} = 0{\cdot}00278$ $pW\,cell^{-1}$;
1 mJ (10^8 cells)$^{-1} \equiv 10$ $pJ\,cell^{-1}$.

incubated with glucose or fructose was greater than that with lactate or β-hydroxybutyrate, except at 20 °C. Energy charge was also higher in cells incubated with the glycolysable substrates. From analyses of end products after the various incubations, the authors calculated total exogenous heat using the stoichiometries of the known pathways. Perhaps surprisingly, they found that heat produced by endogenous metabolism did not vary with temperature or substrate and was comparable to that of sperm incubated without exogenous substrate. Inskeep and Hammerstedt (1985) concluded that 'endogenous metabolism appeared to occur at a constant rate in sperm incubated with different substrates and was independent of the total ATP demand of the cell'. In terms of metabolic regulation, all very puzzling!

It will be remembered that sperm incubated at 35 °C with glucose dissipated considerably more heat than those in the presence of lactate. Inskeep and Hammerstedt (1985) calculated the rate of ATP synthesis for the relevant pathways from the stoichiometry for ATP yield per mol of substrate transformed and gave figures of 9 and 6·2 μmol h^{-1} (10^8 cells)$^{-1}$* for glucose and lactate, respectively. Since, under normal physiological conditions, ATP generation is balanced by consumption, it is necessary to determine the reasons for increased ATP degradation. Apparently sperm motility accounts for 70% of ATP consumed during incubations at 37 °C. Accordingly, the authors measured the velocity of sperm incubated with glucose or lactate but found no difference, indicating that the rates of ATP consumption for motility were identical. Inskeep and Hammerstedt (1985) concluded that the 'extra' ATP synthesized by glucose-consuming cells was expended on reactions associated with substrate cycling of glycolytic intermediates. If this is so, then measurements of oxygen consumption should have revealed differences between the experimental heat equivalent and the theoretical oxycaloric equivalent.

Inskeep *et al* (1985) next turned their attention to differences in cellular metabolism between ejaculated sperm (EJS), which will have interacted with accessory sex gland fluids, and cauda epididymal sperm (CES). Total heat dissipation for the former [-807 ± 98 mJ h^{-1} (10^8 cells)$^{-1}$] was significantly different to that of the latter [-331 ± 35 mJ h^{-1} (10^8 cells)$^{-1}$], when both were incubated with glucose. Under similar conditions, oxygen consumption was threefold higher for EJS [$1{\cdot}49 \pm 0{\cdot}13$ μmol h^{-1} (10^8 cells)$^{-1}$] than for CES ($0{\cdot}55 \pm 0{\cdot}04$). On the other hand, the energy charge was identical in sperm from both sources, indicating that ATP generation and consumption was balanced, as is only right and proper! Inskeep *et al* (1985) calculated that the total rate of ATP synthesis was greater for EJS [13 μmol ATP h^{-1} (10^8 cells)$^{-1}$] than for CES (5 μmol h^{-1}), with less than 2 μmol ATP h^{-1} for cells from each source coming from the degradation of endogenous reserves. Thus, ejaculation would appear to be accompanied by a large increase in the catabolic rate coupled to ergobolic reactions, that is satisfied

*1 μmol h^{-1} (10^8 cells)$^{-1}$ $\equiv 2{\cdot}78 \times 10^{-18}$ mol s^{-1} cell^{-1}.

by the degradation of glucose. The additional ATP generated by ejaculated sperm was partly utilized in increased motility ($85\,\mu m\,s^{-1}$ as against $58\,\mu m\,s^{-1}$ for CES) and possibly also in substrate cycling in the metabolism of glucose at the inter-conversions between glucose/glucose 6-phosphate and fructose 6-phosphate/fructose 1,6-diphosphate. Even so, the authors could not account for all the ATP expenditure by ejaculated sperm. It is interesting to calculate the heat equivalent of oxygen, $\Delta_t Q_{O_2}$, for EJS and arrive at a figure of $-542\,kJ\,mol^{-1}\,O_2$, considerably higher than the average oxycaloric equivalent, $\Delta_k H_{O_2}$, of carbohydrates at $-471\,kJ\,mol^{-1}\,O_2$ (Gnaiger 1983b). Of course, this helps to substantiate the existence of anoxic pathways in ejaculated sperm, as shown by Hammerstedt's group.

Substrate ('futile') cycles loomed large in the assessment by Clark's cohort of metabolic efficiency in isolated rat hepatocytes. Jarrett *et al* (1979) modified an LKB batch microcalorimeter so that the block continually rotated during an experiment and the cell suspension was introduced into and withdrawn from the gold reaction vessels by an injection port secured to the outside of the block. They expressed their findings in terms of the ratio of oxygen uptake (determined polarographically) to heat output but, as the act of an impoverished Nostradamus, the salient ratios are reworked for present purposes as $\Delta_t Q_{O_2}$.

Jarrett *et al* (1979) found that hepatocytes from starved or fed rats in either Krebs-Henseleit bicarbonate buffered physiological saline solution containing gelatin or Eagle's minimum essential medium showed no significant difference in the heat equivalent of oxygen, $\Delta_t Q_{O_2}$ averaging $-542\,kJ\,mol^{-1}$ O_2. Starvation has been shown to decrease the activity of glycolytic enzymes and enhance those of gluconeogenesis but that nutritional state only caused a slight and equivalent increase in both oxygen consumption and heat dissipation. Clearly, starving the animal did not alter the metabolic 'efficiency' of the isolated cells—perhaps one should say 'inefficiency' in the light of the high $\Delta_t Q_{O_2}$ ratios! It follows that the rates of substrate cycling between glucose/glucose 6-phosphate, fructose 6-phosphate/fructose 1,6-diphosphate and pyruvate/phosphoenolpyruvate—the probable reasons for the high $\Delta_t Q_{O_2}$, in addition to the cycles between triacylglycerol/fatty acids, protein/amino acids and glycogen/glucose were not appreciably altered by starvation.

A second factor that could influence the rate of substrate cycles is, of course, substrates. Jarrett *et al* (1979) showed that $10\,mmol\,dm^{-3}$ dihydroxyacetone and $2{\cdot}5\,mmol\,dm^{-3}$ fructose significantly increased heat dissipation without markedly changing the rate of oxygen uptake; thus, $\Delta_t Q_{O_2}$ was raised to -658 and $-649\,kJ\,mol^{-1}\,O_2$, respectively. The authors clearly felt that increased substrate cycling was the cause of these alterations and so incubated cells with $1\,\mu mol\,dm^{-3}$ glucagon, which is known to inhibit metabolic flux though phosphofructokinase and pyruvate kinase (glycolysis) while enhancing flux through fructose 1,6-diphosphatase (gluconeogenesis). There was only a slight and insignificant increase in heat production and oxygen consumption and a small decrease in $\Delta_t Q_{O_2}$ to -526 kJ mol^{-1} O_2. It was concluded that any change in metabolic flux rates was balanced by a decrease in substrate cycling and hence an increase in metabolic efficiency.

Jarrett *et al* (1979) then tried directly to stimulate substrate cycling by incubating cells with an immediate urea-hydrolysing system (10 mmol dm^{-3} NH_4Cl plus 7 μg cm^{-3} urease). Although oxygen uptake changed little, heat output increased significantly leading to an enthalpic change of -680 kJ mol^{-1} O_2, a decrease in metabolic efficiency. In contrast to the work of Sörbris *et al* (1979, 1982) on human white adipocytes, Jarrett *et al* (1979) found no significant difference in the heat output between hepatocytes from rats fed on a normal diet and those on lard-rich or high-sucrose food. The same similarity was also found in comparisons of cellular heat dissipation from homozygous Zucker fa/fa obese rats and their lean littermates.

Continuing their theme of the contribution of substrate cycles to heat 'waste' in hepatocytes, Clark *et al* (1979) made a detailed study of the effects of fructose on intermediary metabolism and the heat equivalent of oxygen. The cells were incubated with 5 mmol dm^{-3} glucose, labelled uniformly with ^{14}C and specifically with ^{3}H at positions 1, 2, 3, or 6, and with 2·5, 7·5 and 25 mmol dm^{-3} fructose. From analysis of the analytical and radioisotope data, the authors conclusively showed that fructose caused increased substrate cycling at glucose/glucose 6-phosphate and fructose 6-phosphate/fructose 1,6-diphosphate. It was possible to calculate that the increase in the former was considerably more than twofold but an exact estimate of the latter could not be made. Incubation of cells with 25 mmol dm^{-3} fructose in the presence of glucose caused decreases of 37% in oxygen consumption and 24% in heat dissipation, resulting in a significant change in $\Delta_t Q_{O_2}$ from -559 to -671 kJ mol^{-1} O_2. Conversely, 2·5 mmol dm^{-3} fructose brought about increases of 43% in oxygen uptake and 64% in heat production, but still a very 'inefficient' $\Delta_t Q_{O_2}$ of -641 kJ mol^{-1} O_2. Clark *et al* (1979) interpreted the latter results, when taken in conjunction with those showing increased formation of $^{3}H_2O$ from (2-^{3}H)- and (3-^{3}H)-glucose, as being due to increased cycling. In this context, it should be noted that, at this physiological concentration of 2·5 mmol dm^{-3}, fructose decreased cellular ATP by 20%. It was more difficult to explain the results with 25 mmol dm^{-3} fructose because, apart from glucose and lactate release, there was a lowering of all other measured parameters including an 80% decrease in the ATP level. This evidence of metabolic stress was attributed to 'excess phosphorylation of fructose'—a problem for ergobolics!

Many biochemists regard ATP generation as normally coupled to consumption by other cellular metabolic pathways. In a few tissues, however, this is not solely the case; for instance, thermogenesis in brown adipose tissue (Nedergaard *et al* 1977). Another candidate tissue for calorigenic processes is the liver and it has long been known that exogenous fatty acids stimulate oxygen uptake considerably more than would be anticipated on the basis of the potential ATP demands arising from the metabolic interactions induced by substrate addition (Berry *et al* 1983). Based on oxygen uptake measurements and analyses of end products, Berry *et al* (1983) inferred that oxidation of acetyl-CoA to CO_2 was preferentially coupled to ATP synthesis, whereas oxidation of fatty acid to acetyl-CoA was obligatorily associated with reversed electron transfer. They stated that ketone body formation 'is not constrained by the hepatic phosphorylation potential and provides a mechanism for the rapid disposal of fatty acids, independent of the ATP demands of the liver'.

Spurred on by these findings, Berry *et al* (1985) recently examined the contribution of hepatic metabolism to diet-induced thermogenesis in rats. These animals should get fat because they live and eat in a delicatessen rather than in a University refectory; but do not, supposedly because of increased thermogenesis. Actually, the process is curiously called 'cafeteria feeding'. Oxygen uptake and heat dissipation of isolated hepatocytes from cafeteria-fed rats were significantly higher than those parameters for cells from pellet-fed (normal diet) animals, although the $\Delta_t Q_{O_2}$ values calculated from the tables of Berry *et al* (1985) were similar, at -506 and -497 kJ mol^{-1} O_2, respectively. Of course, these rates were raised from the levels for endogenous metabolism on addition of exogenous substrates, but there were no significant differences between cells from experimental and control rats when lactate was then added as the substrate. It should be noted, however, that the $\Delta_t Q_{O_2}$ of -537 kJ mol^{-1} O_2 was considerably higher than the $\Delta_k H_{O_2}$ for lactate (-460 kJ mol^{-1} O_2). In contrast to this substrate, glycerol and fatty acids caused a significantly greater heat output and oxygen uptake in cells from cafeteria-fed rats.

Taking these results in conjunction with end product analyses, Berry *et al* (1985) asserted that 'hepatic tissue may be involved to a significant extent in the calorigenic process . . . . . . through the stimulation of specific pathways, namely those associated with the metabolism of glycerol, and the β-oxidation of fatty acids'. In the latter case, Berry *et al* (1985) showed with metabolic inhibitors that most of the reducing equivalents generated during oxidation entered mitochondria beyond the site of rotenone inhibition, but prior to the antimycin-sensitive site—at the level of flavoprotein. Following their earlier suggestion, Berry *et al* (1983) contended that reversed electron transfer, which is an energy-dependent mechanism for driving electrons against the electrochemical potential gradient, is a crucial factor in thermogenesis. It was further suggested that the energy 'conserved' by this process was dissipated as heat by 'redox cycling' exemplified by, for instance, the dihydroxyacetone phosphate and glycerophosphate shuttle between mitochondria and cytosol. It should be emphasized, however, that this discussion was written (Berry *et al* 1985) against a background showing that, for 2 μmol cm^{-3} palmitate at least, metabolism is relatively efficient. Calculations from their data revealed that liver cells from cafeteria-fed and control rats gave $\Delta_t Q_{O_2}$ values of -460

and $-433\,\text{kJ}\,\text{mol}^{-1}\,O_2$, respectively, which is close to the theoretical $\Delta_k H_{O_2}$ for palmitate of $-434\,\text{kJ}\,\text{mol}^{-1}\,O_2$.

As has been stated, in conventional biochemical terms the rate of catabolic ATP generation is balanced by consumption in 'traditional' pathways. While most would agree that (net) ergobolic reactions must form part of the equation, there is more of a conceptual problem in regarding the work performed by muscle (Woledge 1980; Chinet *et al* 1977; Lörinzi 1982) and the cytoskeleton in the movement of non-muscle cells (Kemp 1984) as part of the anabolic half cycle (Gnaiger 1983b). If such an extension can be accepted, however, then it is possible to embrace 'redox cycling' (Berry *et al* 1983, 1985) within the same thermodynamic framework of coupled ATP reactions.

Another candidate for glory or sacrifice in liver thermogenesis is the $(Na^+ + K^+)$-dependent adenosine triphosphatase. Difficulties have arisen because of variations in estimates for energy expenditure by the membrane pump (Clark *et al* 1982) of between 5 and 6% (Chinet *et al* 1977) and 30 and 60% (Ismail-Beigi and Edelman 1970) of basal energy requirements in isolated tissues from euthyroid animals and between 5 and 6% (Folke and Sestoft 1977) and 45 and 90% (Ismail-Beigi and Edelman 1979) in tissues from hyperthyroid animals. Biron *et al* (1979) studied the effects of thyroid hormones on the $(Na^+ + K^+)$-ATPase pump in mammalian skeletal muscle (rat soleus) and found that heat production increased by 30% during the hypothyroid-euthyroid transition and by about 45% in the transition to the hyperthyroid state.

Oubain is a well-documented inhibitor of the plasma membrane $(Na^+ + K^+)$-ATPase and, in fact, has been shown to stop the heat produced by nerves during the falling phase of the action potential, when sodium is actively pumped from nerve fibres (Howarth *et al* 1968; also see review by Richie 1974). Biron *et al* (1979) showed that $1\,\text{mmol}\,\text{dm}^{-3}$ ouabain decreased heat dissipation by 6–8% in muscle from all three groups of rats. In an investigation completed before publication of these results, Clark *et al* (1982) found that $(Na^+ + K^+)$-ATPase activity, heat dissipation and oxygen uptake were increased by 59%, 62% and 75%, respectively, in hepatocytes from triiodothyronine-treated (hyperthyroid) rats. Ouabain decreased oxygen consumption by up to 8% in controls and 15% in cells from experimental animals, while heat production was decreased by 9% in the former and 11% in the latter. In no case was there a significant change in $\Delta_t Q_{O_2}$ values. The authors conclude that 'these results do not support the hypothesis that hepatic $(Na^+ + K^+)$-ATPase plays a major role in increased heat production in hepatocytes from hyperthyroid rats'.

7.5 Conclusions

At this point, I was going to bask in the glory of a vast table citing power values for all things and animals. Of course, it is important to have accurate figures as reference points but their true worth lies in the contribution to our overall understanding of cellular processes. This ambition can only be achieved by thermodynamic and thermochemical analyses, requiring input from classical assays of metabolic pathways and efferent systems. Since, to date, so few effective 'combination' studies have been published, I abandoned the table in favour of a short sermon.

References

Anders A, Welge G, Schaarschmidt B, Lamprecht I and Schaefer H (1977) In: Lamprecht I, Schaarschmidt B (eds) *Application of Calorimetry in Life Sciences*. Berlin, De Gruyter, pp 199–208

Anders A, Schaefer H, Schaarschmidt B and Lamprecht I (1979) *Arch Dermatol Res* **265**, 173–80

Berry M N, Clark D G, Grivell A R and Wallace P G (1983) *Eur J. Biochem* **131**, 205–14

Berry M N, Clark D F, Grivell A R and Wallace P G (1985) *Metabolism* **34**, 141–7

Biron R, Burger A, Chinet A, Clausen T and Dubois-Ferrière R (1979) *J Physiol* (*Lond*) **297**, 47–60

Cerretti D P, Dorsey K J and Bolen D W (1977) *Biochim Biophys Acta* **462**, 748–58

Chinet A, Clausen T and Giradier L (1977) *J Physiol* (*Lond*) **265**, 43–61

Clark D G, Filsell O H and Topping D L (1979) *Biochem J* **184**, 501–7

Clark D G, Brinkman M, Filsell O H, Lewis S J and Berry M N (1982) *Biochem J* **202**, 661–5

Eftimiadi C and Rialdi G (1982) *Cell Biophys* **4**, 231–44

Eftimiadi C and Rialdi G (1984) *J Infect Dis* **150**, 366–71

Eftimiadi C and Rialdi G (1985a) *Thermochim Acta* **85**, 489–92

Eftimiadi C and Rialdi G (1985b) *Microbiologica* **8**, 255–62

Eftimiadi C and Rialdi G (1986) *Microbiologica* (in press)

Flanagan B and Kemp R B (1986) (to be published)

Folke M and Sestoft L (1977) *J Physiol* (*Lond*) **269**, 407–19

Giradier L, Seydoux J, Giacobino J P and Chinet A (1976) In: Jansky L, Musacchia X J (eds) *Regulation of Depressed Metabolism and Thermogenesis*. Springfield, Ill, Thomas, pp 196–212

Gnaiger E (1980) *Thermochim Acta* **40**, 195–223

Gnaiger E (1981) *Thermochim Acta* **49**, 75–85

Gnaiger E (1983a) In: Gnaiger E, Forstner H (eds) *Polarographic Oxygen Sensors*. Berlin and Heidelberg, Springer, pp 337–45

Gnaiger E (1983b) *J Exp Zool* **228**, 471–90

Görman Nordmark M, Laynez J, Schön A, Suurkuust J and Wadsö I (1984) *J Biochem Biophys Meth* **10**, 187–202

Hammerstedt R H and Lovrien R E (1983) *J Exp Zool* **228**, 459–69

Hellgren L and Larsson K (1977) *Experientia* **33**, 97–8

Hoffner S E S, Meredith R W J and Kemp R B (1985) *Cytobios* **42**, 71–80

Howarth J V, Keynes R D and Richie J M (1968) *J Physiol* (*Lond*) **194**, 745–93

Inskeep P B and Hammerstedt R H (1983) *J Biochem Biophys Meth* **7**, 199–210

Inskeep P B and Hammerstedt R H (1985) *J Cell Physiol* **123**, 180–90

Inskeep P B, Magargee S F and Hammerstedt R H (1985) *Arch Biochem Biophys* 241, 1–9

Ismail-Beigi F and Edelman I S (1970) *Proc Natl Acad Sci USA* **67**, 1071–8

Ivlev V G (1934) *Biochem Z* **275**, 49–55

Jarrett I G, Clark D G, Filsell O H, Harvey J W and Clark M G (1979) *Biochem J* **180**, 631–8

Kemp R B (1972) *Science Tools* **19**, 3–5

Kemp R B (1975) *Pestic Sci* **6**, 311–25

Kemp R B (1980) In: Beezer A E (ed) *Biological Microcalorimetry*. New York and London, Academic Press, pp 113–30

Kemp R B (1984) In: Kemp R B Hinchliffe J R (eds) *Matrices and Cell Differentiation*. New York, Alan R. Liss, pp 249–60

Krakauer T and Krakauer H (1976) *Cell Immunol* **26**, 242–53

Leach R (1967) *Ann N Y Acad Sci* **145**, 305–16

Ljungholm K, Wadsö I and Mårdh P-A (1976) *J Gen Microbiol* **96**, 283–8

Ljungholm K, Wadsö I and Kjellen L (1978) *Acta Pathol Microbiol Scand* [*B*] **86**, 121–8

Loike J D, Silverstein S C and Sturtevant J M (1981) *Proc Natl Acad Sci USA* **78**, 5958–62

Lörinzi D (1982) *Studia Biophys* **89**, 81–97

Monti M, Nilsson-Ehle P, Sörbris R and Wadsö I (1980) *Scand J Clin Lab Invest* **40**, 581–7

Nedergaard J, Cannon B and Lindberg O (1977) *Nature* **267**, 518–20

Nicolic D and Neskovic B (1976) *Iugoslav Physiol Pharmacol Acta* **12**, 191–7

Nilsson-Ehle P and Nordin G (1985) *Int J Obesity* **9**, Suppl. 1, 169–72

Pätel M (1981) *Thermochim Acta* **49**, 123–9

Pätel M, Reichert U, Schaarschmidt B and Lamprecht I (1980) In: Hemminger W (ed) *Thermal Analysis*, vol II. Basel, Birkhäuser, pp 559–64

Pätel M, Schaarschmidt B and Reichert U (1981) *Br J Dermatol* **105**, Suppl. 20, 60–1

Pätel M, Schaarschmidt B and Lamprecht I (1982) In: Miller B (ed) *Thermal Analysis*, vol II. Chichester, Wiley, pp 857–62

Reichert U and Schaarschmidt B (1986) *Experientia* **42**, 173–4

Richie J M (1974) *Prog Biophys Mol Biol* **26**, 147–88

Schaarschmidt B and Reichert U (1981) *Exp Cell Res* **131**, 480–3

Schaarschmidt B and Reichert U (1986) *Thermochim Acta* (in press)

Sörbris R, Nilsson-Ehle P, Monti M and Wadsö I (1979) *FEBS Lett* **101**, 411–4

Sörbris R, Monti M, Nilsson-Ehle P and Wadsö I (1982) *Metabolism* **31**, 973–8

Sörbris R, Olsson S A, Monti M, and Nilsson-Ehle P (1986) *Clin. Nutrition* **6**, 195–7

Spink C and Wadsö I (1976) In: Glick D (ed) *Methods of Biochemical Analysis*, vol 23. New York, Wiley, pp 1–158

Suurkuusk J and Wadsö I (1982) *Chemica Scripta* **20**, 155–63

Thorèn S and Holma B (1986) (to be published)

Valdemarsson S, Fagher B, Hedner P, Monti M and Nilsson-Ehle P (1986) *Acta Endocrinol* **108**, 361–6

Wadsö I (1976) *Biochem Soc Trans* **4**, 561–6

Woledge R C (1980) In: Beezer A E (ed) *Biological Microcalorimetry*. New York and London, Academic Press, pp 145–62

Chapter 8

Transfer Thermodynamics

A. E. Beezer, R. J. Miles and B. F. Perry

Nomenclature

ΔG	Gibbs Energy change: specified as appropriate for the process discussed as $\Delta_{trs}G$ for a solute transfer: $\Delta_{sln}G$ for a solution process
ΔH	enthalpy change, specified as $\Delta_{trs}H$
ΔS	entropy change, specified as $\Delta_{trs}S$
$\Delta\mu_L$	change in chemical potential upon transfer of a lipophilic group (e.g. CH_2) from water to a non-aqueous solvent
$\Delta\mu_H$	change in chemical potential upon transfer of a hydrophobic group: arbitrary assignment of hydrophilicity to the parent group of a molecule, i.e. the molecule minus R as in

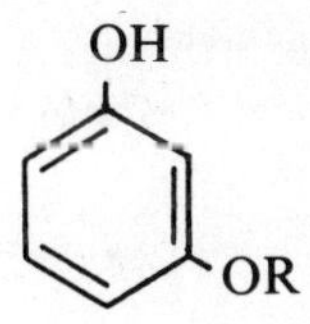

c	concentration of a solute; subscript (o) or (w) denotes in organic solvent or water
P	partition coefficient, defined as: $P = c_{(o)}/c_{(w)}$
o	octanol
pc	propylene carbonate
h	heptane
n	number of lipophilic groups (CH_2) in the R grouping of a molecule
$V_{(w)}$, $V_{(o)}$	molar volumes of water and organic solvent under study
(I) and (II)	as suffixes refer to system (I) as water and organic solvent and system (II) as water and some other organic solvent

8.1 Introduction

This account of transfer thermodynamics makes no attempt to be comprehensive. It simply concentrates on some thermodynamic and *calorimetric* studies related to the process: drug in aqueous solution → drug in lipid-like solvent/membrane. More comprehensive accounts of the pharmaceutical significance of transfer processes can be found in a review by Tomlinson (1983), and in less detail in books (Taylor and Kennewall 1981; Smith and Williams 1983; Nogrady 1985).

8.2 Background

Drugs administered to humans pass from the point of entry (oral, sublingual, rectal, parenteral intravenous, intramuscular, subcutaneous, pulmonary, topical, etc.) to the site of action; an area described as the pharmacokinetic phase. The passage from the site of entry to the site of action will involve the transfer of the drug from one medium to another and will involve the movement of the drug across tissue membranes and microbial membranes (Albert 1979; Taylor and Kennewall 1981; Smith and Williams 1983; Nogrady 1985). In attempting to discover the most important step in this process, the use of partition constants has found much favour. It is surprising that dynamic processes have found some order through the use of equilibrium expressions.

The basis of the interest in partition coefficients and their quantitative relationship with biological activity is the development of the Hammett equation (Wells 1968; Johnson 1973). This equation attempts to quantify the effects of substituent groups on molecular systems. Thus the electronic and steric effects of substitution which can be measured in one system are extrapolated to other molecular systems. It is this capacity for prediction which is at the heart of Quantitative Structure Activity Relationships (QSAR), i.e. the attempt to predict Biological Response from some easily measured property of the drug.

The Hammett equation is, in general form:

$$\log \frac{K}{K_0} = \sigma\rho \tag{8.1}$$

where K is an equilibrium or rate constant for some substituent group X and K_0 is that for the 'parent' structure, i.e. X = H; σ is a constant characteristic of the electronic effect of X and ρ is termed the reaction constant. Since for each constant in Equation (8.1) it is possible to write an equation of the form:

$$\Delta G^{\ominus} = -RT \ln K \tag{8.2}$$

then relationships such as Equation (8.1) are known as Linear Free Energy

Relationships (LFER). The direct application of relationships such as described by Equation (8.1) was not particularly appropriate to systems involving living species, i.e. *in vivo* systems in contrast to the *in vitro* systems described by the Hammett equation (8.1).

Hansch modified the Hammett approach and has developed equations of the form of (8.3):

$$\log \frac{1}{c} = k_1 \log P + k_2\sigma + k_3 \tag{8.3}$$

where: c is the concentration of drug required to elicit a specific effect; P is the partition coefficient (to be specified below) and k_1, k_2 and k_3 are constants (Leo *et al* 1971; Hansch and Leo 1979).

The partition coefficient is defined as for ideal solutions as:

$$P = \frac{x_{(o)}}{x_{(w)}} \tag{8.4}$$

where $x_{(w)}$ and $x_{(o)}$ are the mole fractions of solute in an aqueous (w) and non-aqueous phase (o) which are in equilibrium and in contact with each other. In practice Equation (8.4) is often written as:

$$P = \frac{c_{(o)}}{c_{(w)}} \tag{8.5}$$

where the c's are some other concentration measure (mol dm^{-3}, molal) with the proviso that the concentrations in each solvent system are within the range where Henry's Law is obeyed (Cratin 1968; De Lisi *et al* 1980).

8.3 Thermodynamic Aspects

Cratin (1968) has described how P and the derived Gibbs energies [e.g. from equations such as Equation (8.2)] may be analysed for homologous series to yield the desired objective of group contributions to Gibbs energies and hence to values of P for use in equations of the type of Equation (8.3). The analysis as modified by Beezer *et al* (1986a) shows that:

$$P = -\frac{n\Delta\mu_{\mathrm{L}}^{\ominus}}{RT} - \frac{\Delta\mu_{\mathrm{H}}^{\ominus}}{RT} + \ln V_{(w)}/V_{(o)} \tag{8.6}$$

where n is the number of lipophilic groups present in a molecule (e.g. methylene groups) each with a contribution to the overall change in chemical potential upon transfer of the group from water to a non-aqueous solvent of

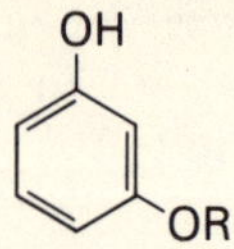

where R = Me, Et, Pr, ...
n = 1, 2, 3, ...

and

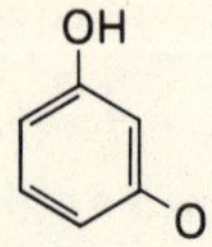

represents parent group (see Table 8.5 and text).

Figure 8.1.

$\Delta\mu_L^{\ominus}$: $\Delta\mu_H^{\ominus}$ is the change in chemical potential upon transfer of the 'parent' structural fragment of the molecule (Figure 8.1), designated here as hydrophilic, from water to the same non-aqueous solvent; $V_{(o)}$, $V_{(w)}$ are the molar volumes of the two solvents studied. Inspection of Equation (8.6) shows that plots of ln P against n should be linear of slope $-\Delta\mu_L^{\ominus}/RT$ and intercept $-\Delta\mu_H^{\ominus}/RT + \ln V_{(w)}/V_{(o)}$. Thus if such linearity in plots of ln P against n exists then measurement of P for a few members of a series would allow prediction of the values of P for higher members; eventually through Equation (8.3) to the establishment of which homologue *may* show greatest use in application.

The discussion above has not addressed the issues of which solvent to use in partition studies and what standard states are appropriate in the thermodynamic analysis. The selection of the solvent system and the choice of standard states are, however, closely related.

Inspection of the literature shows that a wide choice of standard states has been made to define the standard state Gibbs energies and entropies of solution and to express solubilities of solutes in liquids. Abraham (1984) has discussed the quantitative effects on $\Delta_{sln}G^{\ominus}$ for various choices of standard states (Table 8.1). The table indicates that, whilst comparison of individual values of $\Delta_{sln}G^{\ominus}$ is inappropriate where different standard states have been employed, differences, i.e. $\Delta(\Delta_{sln}G^{\ominus})$, are independent of the choice of standard states.

Solution processes are likewise subject to similar confusion in that, again, many and varied standard states have been used (Beezer *et al* 1983a,b) and advocated (Tanford 1980). The standard state adopted (Tanford 1980) by

Table 8.1 Correction parameters for Standard State difference

Standard state		$\Delta_{sln}G^{\ominus}$/kJ mol^{-1}		
Gas	*Solvent*	*Water*	*Methanol*	*Hexane*
1 atm	unit mol fraction	0	0	0
1 atm	1 mol dm^{-3}	9·95	7·93	5·02
1 atm	1 mol kg^{-1}	9·96	8·53	6·06
1 mol dm^{-3}	1 mol dm^{-3}	17·87	15·86	12·95
1 kPa	1 mol dm^{-3}	15·62	13·61	10·70
	solvent relative molecular mass	18·015	32·04	86·18
	solvent density /g cm^{-3}	0·9971	0·7865	0·6548

defined so that

$$\Delta_{sln}G^{\ominus}\ (1\ \text{atm} \rightarrow \text{unit mol fraction}) = \Delta_{sln}G^{\ominus}\ (\text{any other standard state}) + \Delta_{sln}G^{\ominus}$$

some is the pure liquid solute and the ideal solution (unit mol fraction of solute). The scheme described below indicates the relationships involved:

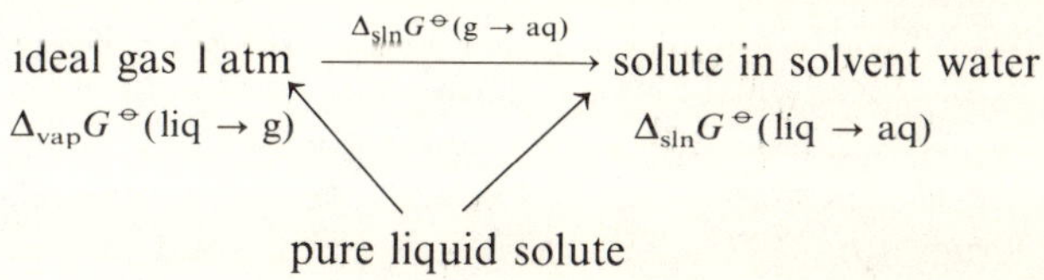

and hence:

$$\Delta_{sln}G^{\ominus}(g \rightarrow aq) + \Delta_{vap}G^{\ominus}(liq \rightarrow g) = \Delta_{sln}G^{\ominus}(liq \rightarrow aq).$$

$\Delta_{vap}G^{\ominus}$ and other thermodyamic parameters $\Delta_{vap}H^{\ominus}$ and $\Delta_{vap}S^{\ominus}$ will differ from solute to solute. Moreover, the conversion $\Delta_{sln}G^{\ominus}(liq \rightarrow aq)$ to $\Delta_{sln}G^{\ominus}(g \rightarrow aq)$ will differ for all solutes no matter how closely related. Abraham (1979) has underlined the fact that choice of such a standard state assigns a different standard state to each solute. Ideally, the standard state selected should not include considerations of solute-solute interactions in the liquid phase: this can be achieved by selection of the gas phase at unit atmosphere as the standard state. Under those conditions discussion of $\Delta_{sln}G^{\ominus}(g \rightarrow sln)$ would be interpreted through consideration of only solute-solvent interactions.

However, it has been argued that for consideration of drug/solution thermodynamics the selection of the vapour phase as the standard state is inappropriate. It is asserted (Rytting *et al* 1972), that the properties of the dilute solution are important but the properties of the gas phase are not particularly relevant. Furthermore, the conformation of moderate to long chain molecules may be quite different in the vapour phase from their conformation in solution phase; this then may make the interpretation of experimental data very much more complex. Rytting *et al* (1972) have thus proposed that the universal standard state for drug molecules be the hypothetical 1 molal, mol dm^{-3} or unit mol fraction solution acting as if it were infinitely dilute (where the solvent is a suitable aliphatic hydrocarbon such as cyclohexane or iso-octane). Water, because of its highly complex and structured nature which leads to significant solute-solvent and solvent-solvent interactions, is not here (Rytting *et al* 1972) regarded as the best common solvent.

Values of $\Delta_{\text{sln}}G^{\ominus}$ can be determined from Henry's Law constants or from solubilities, whereas the corresponding enthalpies may be determined from the van't Hoff isochore or from direct calorimetry. Biltonen (1980) has suggested that, for equilibria involving macromolecules it is unlikely that K can be known to be better than $\pm 30\%$. For acid-base equilibria King (1965) has calculated the effect errors in the determination of K have upon the derived values of $\Delta H^{\ominus}$, $\Delta S^{\ominus}$ and $\Delta C_p^{\ominus}$. The extract from King's table (Table 8.2) shows that calorimetric data is to be preferred. This is more true for systems involving macromolecules and for sparingly soluble substances, i.e. for drug-based systems in the context of this contribution. Tomlinson (1983) has described and reviewed in great detail the application of LFERs in pharmaceutical, biochemical and biological systems. Attention was directed in this review to the statistical basis of the results used to correlate ΔH with ΔS and hence to determination of isokinetic-isoequilibrium relationships. Following Krug *et al* (1976a,b,c, 1977) it is emphasized that in order for thermodynamic data to show 'chemical causality' then it is $\Delta G^{\ominus}$ and $\Delta H^{\ominus}$ data that must be linearly related at the harmonic mean of the experimental temperatures.

In practical terms the parameters of interest for drug related studies are the enthalpy, entropy and Gibbs energy for the transfer process from water to

Table 8.2 Propagation of errors in $\Delta H^{\ominus}$, $\Delta S^{\ominus}$ and $\Delta C_p^{\ominus}$ at 298 K.

Conditions of experiment

T range/K	*293–303*		*273–323*		*278–323*	
T interval/K	*2*		*10*		*5*	
	pK ± *0·02*	*pK* ± *0·001*	*pK* ± *0·02*	*pK* ± *0·001*	*pK* ± *0·02*	*pK* ± *0·001*
SD in $\Delta H^{\ominus}$ /kJ mol^{-1}	±4·06	±0·21	±0·84	±0·04	±0·79	±0·04
SD in $\Delta S^{\ominus}$ /J mol^{-1} K^{-1}	±13·81	±0·67	±2·72	±0·14	±1·33	±0·07
SD in $\Delta C_p^{\ominus}$ /J mol^{-1} K^{-1}	±2768	±138·1	±113·0	±5·44	±117·2	±5·86

some suitable lipoid phase. Such measurements are conveniently made by (*a*) determination of $\Delta_{trs}G$ from measurements of P values and application of Equation (8.2), (*b*) determination of $\Delta_{trs}H$ from either direct calorimetry or from the van't Hoff isochore, and (*c*) $\Delta_{trs}S$ from:

$$\Delta_{trs}G = \Delta_{trs}H - T\Delta_{trs}S \tag{8.7}$$

As has been argued above, direct calorimetric methods for the determination of $\Delta_{trs}H$ are to be preferred. There are two main experimental approaches to the determination of $\Delta_{trs}H$, one being direct measurement and the other determination of $\Delta_{sln}H$ for the solute in each solvent and then calculation of $\Delta_{sln}H$ from:

$$\Delta_{trs}H = \Delta_{sln}H_{(\text{lipoid solvent})} - \Delta_{sln}H_{(\text{water})} \tag{8.8}$$

Direct measurement has been proposed by several workers (Breslauer *et al* 1974, 1978; De Lisi *et al* 1980; Goffredi and Liveri 1981; Tomlinson 1983) and each method has some practical disadvantages, e.g. lengthy experimental times, separate establishment of extent of transfer, etc.but all are claimed to yield satisfactory results. Determination of $\Delta_{sln}H$ data has been favoured by other authors (Beezer *et al* 1983a,b, 1986a) but it is again lengthy. Measurement of $\Delta_{sln}H$ does, however, have a significant advantage over direct determination of $\Delta_{trs}H$ in that any anomalous results can be indicated as arising from behaviour in one solvent system and hence may give more insight into the solution/transfer process.

Abraham (1984) has discussed the thermodynamics of solution for many solutes in water and has shown that for most homologous series the parameters for the process gas to aqueous phase are linear in the number of carbon atoms in the solute. From such linear relationships, methylene and group contributions may be obtained. Such contributions are, however, not constant but vary from homologous series to homologous series. Indeed, within some homologous series (e.g. the alkan-1-ols and *n*-alkanes) the methylene group contributions are not constant. Within this analysis Abraham (1979) has attempted a quantitative assessment of the hydrophobic effect (Tanford 1980). Analysis of $\Delta_{sln}G$ data (at 298·15 K) showed that solution of alkanes but not rare gases in water is quite anomalous. The methylene group contribution for partition of hexane, for example, between hexane and water is 3·85 kJ mol^{-1} in favour of hexane, and this can be separated (Abraham 1979) into a favourable gas → hexane contribution of 3·10 kJ mol^{-1} and an unfavourable gas → water contribution of 0·75 kJ mol^{-1}. This latter quantity is shown to result from a 'true' hydrophobic contribution of 2·26 kJ mol^{-1} and a favourable normal solvent effect of 1·51 kJ mol^{-1}.

The partition process referred to is:

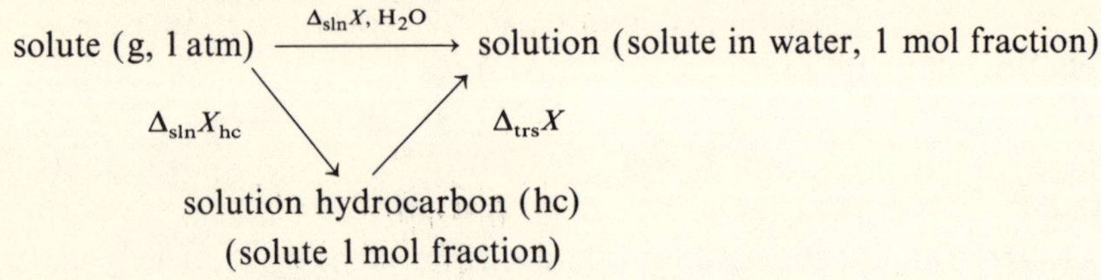

where

$$\Delta_{trs}X = \Delta_{sln,hc}X - \Delta_{sln,H_2O}X$$

Therefore whilst $\Delta_{trs}X$ ($X = G, H, S$) is desired for use in QSAR it is more instructive to investigate $\Delta_{sln}X$ in each selected solvent system. However, such detailed investigations are rare.

In an extensive study of a series of *m*-alkoxy phenols the values for $\Delta_{trs}X$ and $\Delta_{sln}X$ have been determined for the following solvent systems: water and propylene carbonate, heptane, octan-1-ol. It was found that in the water/octan-1-ol system, values of $\Delta(\Delta_{trs}G)$ were not constant but showed oscillatory behaviour (Table 8.3). It thus appears that there exist two separate linear relationships; one for the odd carbon number members of the series and one for the even carbon number members. Measurement of $\Delta_{sln}G$ in water indicated that it was the properties of the water solution that dictated the oscillatory nature of the results for $\Delta_{trs}G$. The oscillation presumably has its origins in the packing of the solute into the solvent (water) and hence to differences in solute-solute and solute-solvent interactions. Indeed, data in the literature (Aveyard and Mitchell 1969, 1970) for partitioning of alkanols and alkanoic acids appears to display similar valued oscillations in $\Delta(\Delta_{trs}G)$. The existence of such oscillations suggests that the simple linearity between $\Delta_{trs}G$ and the number of carbon atoms present in the solute may not always be true.

Table 8.3 Values of $\Delta_{trs}G$ and $\Delta(\Delta_{trs}G)$ for transfer of resorcinol monoethers from water to octan-1-ol

Phenol	$-\Delta_{trs}G$	$\Delta(\Delta_{trs}G)$	$\Delta_{sln.w}G$	$\Delta(\Delta_{sln.w}G)$	$\Delta_{sln,w/o}G$	$\Delta(\Delta_{sln,w/o}G)$
m-methoxy	9·42		13·08		13·72	
m-ethoxy	11·77	2·35	15·93	2·85	16·47	2·75
m-propoxy	15·23	3·46	19·33	3·40	20·16	3·69
m-butoxy	17·78	2·55	22·22	2·89	23·28	3·12
m-pentoxy	21·19	3·41	25·63	3·41	27·02	3·74

Values of $\Delta_{sln}G$ for solution of resorcinol monoethers in water ($\Delta_{sln.w}G$) and in octan-1-ol saturated water ($\Delta_{sln,w/o}G$), together with the respective $\Delta(\Delta_{sln}G)$ values. All values are in kJ mol^{-1}

Table 8.4 Values of $-\Delta\mu_L^\ominus$, $-\Delta\mu_H^\ominus$ derived from plots described by Equation (8.6). Values in $kJ\,mol^{-1}$. Values relate to solute transfer from water to described non-aqueous solvent[a]

Temperature K	$-\Delta\mu_L^\ominus$	$-\Delta\mu_H^\ominus$	*Solvent*
283·15	3·09	−4·08	h
	0·81	11·28	pc
288·15	3·15	−3·95	h
	1·01	11·06	pc
	2·83	9·78	o
293·15	3·09	−3·25	h
	1·12	10·72	pc
	2·90	9·76	o
298·15	3·07	−2·92	h
	1·11	10·69	pc
	2·96	9·75	o
303·15	3·16	−2·68	h
	1·11	10·57	pc
	3·03	9·78	o
308·15	3·32	−2·88	h
	1·17	10·26	pc
	3·09	9·82	o
313·15	3·10	−1·79	h
	1·13	10·07	pc

[a]pc, water-saturated propylene carbonate; o, water-saturated octan-1-ol; h, water-saturated heptane.

A more detailed analysis of the $\Delta_{trs}G$ data is presented in Table 8.4 (Beezer *et al* 1986a) where values for $\Delta\mu_L^\ominus$ and $\Delta\mu_H^\ominus$ are displayed [values derived from application of Equation (8.6)]. The values illustrate a proper conformity with the apparent hydrophobicity of the non-aqueous solvent system (water is present in both octanol and in propylene carbonate in the partition experiment). A notable feature of Table 8.4 is that $\Delta\mu_H^\ominus$ shows a dependence upon temperature whereas $\Delta\mu_L^\ominus$ appears to be independent of temperature. Plots of $\Delta\mu_H^\ominus/RT$ against $1/T$ allow derivation of $\Delta_{trs}H_H$ and hence of $\Delta_{trs}S_H$ and these values, together with those for transfer of the lipophilic methylene groups to each solvent system, are displayed in Table 8.5. A more limited study by Lissi (1981) for partitioning of alkan-1-ols between heptane and water also implies that $\Delta_{trs}H_L$ is zero.

The importance of the $\Delta_{trs}S$ term in determining values of $\Delta_{trs}G$ is apparent, and not surprising. However, the values of $\Delta_{trs}X$ reported in Table 8.5 are for transfer of solute from bulk water solvent to bulk 'lipoid' solvent. Whilst bulk water solvent may be satisfactory for representation of one biological phase it is clear that bulk, say octan-1-ol, does not represent the ordered lipid structure of the cell membrane which it is supposed to mimic. Thus a means

Table 8.5 Thermodynamic data for transfer of the parent structure and for CH_2 group from water to the stated solvent at 298·15 K

Solvent	$\Delta_{trs}H$/kJ mol^{-1}	$\Delta_{trs}G$/kJ mol^{-1}	$\Delta_{trs}S$/kJ mol^{-1} K^{-1}
parent structure			
heptane	23·5	2·92	69·0
propylene carbonate	– 22·1	– 10·69	– 38·3
octan-1-ol	– 9·2	– 9·75	1·8
CH_2			
heptane	0	– 3·07	13·3
propylene carbonate	0	– 1·11	3·7
octan-1-ol	0	– 2·96	9·9

of establishing a scale for relating properties of non-aqueous solvents would be extremely useful.

Collander (1951) noted that, on an empirical basis, partition coefficients for the same solute transferred between water and solvent I and between water and solvent II were related through:

$$\ln P(\mathrm{I}) = a + b \ln P(\mathrm{II}) \tag{8.9}$$

This equation, whilst empirical, has been invested with meaning by many workers. For example, Katz and Diamond (1974a,b) suggested that the value of b reflected the similarity of the solvent environment with respect to the solute and referred to b as a selectivity constant. Leo *et al* (1971) describe b as a measure of the sensitivity of the system to perturbation by hydrophobic effects and the value of a is said to reflect 'the sensitivity of the biochemical system and the intrinsic activity of a set of congeners.' However substitution into Equation (8.9) of equations written for solvent I and for solvent II of the form of Equation (8.6) produces an equation (Beezer *et al* 1986a) from which thermodynamic values can be derived for both a and b. It can be shown (Beezer *et al* 1986a) that a has no physical or chemical meaning. Indeed, this is apparent from Equation (8.9) where upon setting $\ln P(\mathrm{II}) = 0$, i.e. $P(\mathrm{II}) = 1$ then $\ln P(\mathrm{I}) = a$; $\ln P(\mathrm{II})$ can only be made equal to zero by manipulation of the value of n since all other parameters have, if an LFER holds, constant values. For the system for which most data is available (m-alkoxy phenols and the solvent systems described for these solutes in Table 8.4) it can be shown that when solvent I is heptane and solvent II is propylene carbonate a value of $n = -6{\cdot}15$ is required to set $\ln P(\mathrm{II}) = 0$. A negative value of n has no meaning. However, consideration of b shows that it is represented by:

$$\frac{\Delta\mu_{\mathrm{L}}(\mathrm{I})}{\Delta\mu_{\mathrm{L}}(\mathrm{II})}$$

Table 8.6 Values of b calculated as described in the text and as derived from plots of $\ln P(\mathrm{I})$ against $\ln P(\mathrm{II})$

Solvent		*b*	
(I)	*(II)*	*Calc.*	*Expt.*
h	pc[a]	2·93	2·76
pc	o	0·36	0·37
h	o	1·06	1·04

[a]pc, water-saturated propylene carbonate; o, water-saturated octan-1-ol; h, water-saturated heptane.

Thus b has some thermodynamic significance and if some suitable solvent is selected as solvent II then all other solvents may be evaluated as partitioning solvents with respect to this solvent through use of values of b. Moreover, as was noted earlier, values of $\Delta\mu_{\mathrm{L}}$ are insensitive to temperature (Table 8.4) hence all plots of $\ln P(\mathrm{I})$ against $\ln P(\mathrm{II})$ at a series of temperatures should be parallel. This is indeed observed for the system discussed here: the values of b are displayed in Table 8.6. Hence b values may be used to allow identification of a bulk solvent which more satisfactorily represents the properties of biological membranes; this would be of undoubted advantage in QSAR studies.

Values of b may be made more useful perhaps if one system, e.g. water/octan-1-ol, was selected as a reference system and all values of b were recorded relative to this system. Octan-1-ol may be useful here, since it is for partition in the water/octan-1-ol system that most use has been found in the establishment of equations of the type; Equation (8.3).

A first approximation to the membrane of an organism may be liposomes; these are lipid bi- or multi-lamellar vesicles which necessarily are more ordered than bulk solvents. Not surprisingly, the thermodynamic parameters for partitioning of solutes into liposomes from water have been studied; although always through use of the van't Hoff isochore (Tomlinson 1983; Saket *et al* 1984; Arrowsmith *et al* 1983). Saket *et al* (1984) studied the partitioning of some 21-alkyl esters of hydrocortisone and cortisone between water and isopropyl myristate (IPM), dimyristoyl phosphatidylcholine (above and below the transition temperature, T_{c}) and report data on partitioning of these same solutes between water and diethyl ether and octan-1-ol. Their data can be examined via the Collander equation as described above. The results of this analysis are described here in Table 8.7. The values of b calculated from the plots of $\ln P(\mathrm{I})$ against $\ln P(\mathrm{II})$, again selecting octan-1-ol as the 'standard' reference system, are in excellent agreement with the values calculated from:

$$b = \frac{\Delta\mu_{\mathrm{L}}(\mathrm{I})}{\Delta\mu_{\mathrm{L}}(\mathrm{II})}$$

Table 8.7 Values of b calculated from plots of $\ln k$ (I) against $\ln k$ (II) (slope) and from $b = \Delta\mu_L(\text{I})/\Delta\mu_L(\text{II})$. Data recalculated from Saket *et al* (1984) taking system II as water/octan-1-ol (see text)

	Solvent system I: water/							
	diethyl ether		*IPM*		*liposome* ($T < T_c$)		*liposome* ($T > T_c$)	
	slope	*b*	*slope*	*b*	*slope*	*b*	*slope*	*b*
hydrocortisones	1·895	1·53	0·655	0·625	0·584	0·545	0·655	0·616
cortisones	1·828	1·30	0·589	0·582	0·708	0·672	0·589	0·744

Of course the values of b confirm the conclusions of Saket *et al* (1984) that IPM more closely corresponds with the behaviour of liposomes (both above and below T_c) than does the behaviour of diethyl ether to represent a lipoid solvent. There is, unfortunately, little other extensive data that allows of such a detailed examination.

The final objective is the measurement of those thermodynamic parameters for the interaction of aqueous drug solutions with biological cell systems. Tomlinson (1983), in drawing attention to the proposed use of liposomes as models of biological membranes, notes that there is little data available to test this hypothesis. Indeed, only results on non-electrolyte permeation across red cell membranes (Good 1967; Galey *et al* 1973) and for non-electrolyte flux through the toad urinary bladder (Bindslev and Wright 1976) are cited by Tomlinson (1983).

A possible test for such modelling success could be the direct measurement of $\Delta_{trs}H$ for solute transfer from water (or isotonic aqueous solution) to liposomes and to biological cells. Tomlinson (1983) notes that 'The use of batch microcalorimetry in any further studies on liposome energies is seemingly highly indicated', whereas Saket *et al* (1984) remarks that 'Although microcalorimetry may be employed to determine ΔH directly, such techniques are not applicable to structured phases such as liposomal membranes'. However, these authors believe that direct microcalorimetric measures of $\Delta_{trs}H$ for solute transfer from aqueous solution to liposomes would be possible, and very advantageous. In particular, the difficulties associated with the variation in the properties of liposomes (fluidity, volume, enclosed water, etc.) as a function of temperature will be avoided. Of course, avoidance of the error propagation problems associated with van't Hoff techniques referred to earlier would be entirely absent too.

An attempt to measure directly $\Delta_{trs}H$ for transfer of some *m*-alkoxy phenols from isotonic solution to *Escherichia coli* cells has been reported by Beezer *et al* (1986b). The solutes (pure liquids) were added to the bacterial cells suspended in 25% strength Ringer solution. After correction for enthalpy of

Table 8.8 Values of $\Delta_{trs}H$ for transfer of *m*-alkoxyphenol from water to cells, octan-1-ol, heptane and propylene carbonate

Solute *substituted phenols*	$\Delta_{trs}H/\text{kJ mol}^{-1}$			
	Cells	*Octan-1-ol*	*Heptane*	*Propylene carbonate*
m-methoxy-	−0·22	−8·03	20·9	23·2
m-ethoxy-	−1·1	−6·95	19·3	23·4
m-propoxy-	−2·02	−6·96	16·0	23·9
m-butoxy-	−4·06		13·9	23·4
m-pentoxy-	−5·14		12·0	23·2

dilution of the solutes the values of $\Delta_{trs}H$ derived are as shown in Table 8.8. These values were calculated from the experimental data on the assumption that the first, rapid event corresponds to dilution and transfer and that the subsequent, very lengthy thermal events were associated with the *biological* consequences of the interaction. Given this assumption it is notable, from Table 8.8 where $\Delta_{trs}H$ data are compared for several solvent systems, that values for transfer to cells are very different from the values for the bulk solvent systems. These data suggest that bulk solvent systems do not represent very well the behaviour of organized 'solvent' systems such as biological membranes.

An alternative approach to QSAR attempts to relate biological activity to structure rather than relating some physical property, e.g. partition coefficient, to structure. Conventionally described as the Free-Wilson Method the approach assumes (Chu 1980) that biological activity can be additively calculated from substituent constants, e.g.:

$$\log\frac{1}{c} = \sum a_i X_i + \mu_0$$

where $\mu_0 = 1/c$ for the parent compound, a_i the contribution of the ith substituent, X is 1 if the group is present and zero if the group is absent in the studied molecule; as in Equation (8.3) c is the concentration of the drug required to elicit a standard response.

Frequently difficulties are encountered in the evaluation of a 'standard response' (for a review of analytical methods see for example Kavanagh 1963). Recently, however, flow microcalorimetry has been advanced (Chowdhry *et al* 1983) as a sensitive, reproducible technique for the quantitative determination of drug bioactivities upon interaction of the drugs with metabolizing organisms. The technique requires only low densities of microorganisms (10^3–10^4 cells cm^{-3}) and is relatively rapid (a single assay requiring from 40 min to about 5 h). The experimental product, a linear relationship between log (dose) and response (defined in terms which may be arbitrary or may be expressed in time or μW) is that normally found in drug analysis. The calculation of the parameters of this line (slope and intercept) allow, from values of the intercept, the definition of a term log $(\text{dose})_{max}$. Log $(\text{dose})_{max}$ is, then, the logarithm of the maximum concentration of the drug which can be applied without eliciting a response from the sensitive, interacting organism. In the limited studies made so far (Beezer *et al* 1986c) two series of drugs have been found to yield linear plots of log $(\text{dose})_{max}$ against carbon number in the side chain (Figure 8.1). Thus the slope of this line represents the increment in log $(\text{dose})_{max}$ per CH_2 group and the intercept represents the log $(\text{dose})_{max}$ for the parent molecule in the absence of any group at all in place of R (termed here parent group). If this relationship is shown to hold for other series, then it may be possible to calculate the value of *c* (the concentration of drug that can be added before eliciting a response from the sensitive, responding organism) from group contributions. This concentration may be useful, since it defines a concentration (i.e. solubility) which must be *exceeded* for the drug to have useful biological activity. Such relationships in relation to solubility studies and their relationship with structure may prove rather rewarding.

The conclusions of this intentionally non-comprehensive study of QSAR and the thermodynamics of drug partitioning has highlighted the need for better and directly measured values of $\Delta_{trs}H$ to relate to $\Delta_{trs}G$ and hence to the determination of $\Delta_{trs}S$. The insight into interactions which these parameters will give for systems involving liposomes and, possibly, whole cellular systems will be profound. The need, too, has been emphasized for the selection of manageable bulk systems which adequately mimic biological cells. In both this pursuit and in the pursuit of quantitative measures of biological activity the role of microcalorimetry, it is suggested, is profound.

References

Abraham M H (1979) *J Am Chem Soc* **101**, 5477–84

Abraham M H (1984) *J Chem Soc Faraday I* **80**, 153–82

Albert A (1979) *Selective Toxicity*, edn VI. London, Chapman and Hall

Arrowsmith M, Hadgraft J and Kellaway I W (1983) *Biochim Biophys Acta* **750**, 149–56

Aveyard R and Mitchell R W (1969) *Trans Faraday Soc* 2645–53

Aveyard R and Mitchell R W (1970) *Trans Faraday Soc* 37–42

Beezer A E, Hunter W H and Storey D E (1983a) *J Pharm Pharmacol* **35**, 350–7

Beezer A E, Hunter W H and Storey D E (1983b) *J Med Chem* **26**, 757–9

Beezer A E, Volpe P L O and Hunter W H (1986a) *J Chem Soc Faraday I* **82**, 2863–71

Beezer A E, Volpe P L O and Hunter W H (1986b) *J Chem Soc Faraday I* **82**, 2929–32

Beezer A E, Volpe P L O, Gooch C A, Hunter W H and Miles R J (1986) *Int J Pharm* **29**, 237–42

Biltonen R L (1980) In: Beezer A E (ed) *Biological Microcalorimetry*. London, Academic Press
Bindslev N and Wright E M (1976) *J Membr Biol* **29**, 265–88
Breslauer K J, Terrin B and Sturtevant J M (1974) *J Phys Chem* **78**, 2363–9
Breslauer K J, Witkowski L and Bulas K (1978) **82**, 675
Chowdhry B Z, Beezer A E and Greenhow E J (1983) *Talanta* **30**, 209–43
Chu K C (1980) In: Wolff M E (ed) *The Basis of Medicinal Chemistry*, Part 1. New York, Wiley-Interscience, pp 393–418
Collander E A (1951) *Acta Chem Scand* **5**, 774–80
Cratin P D (1968) *Ind Eng Chem* **60**, 14–19
de Lisi R, Goffredi M and Liveri V T (1980) *J Chem Soc Chem Comm* **8**, 380–1
Galey W R, Owen J D and Solomon A K (1973) *J Gen Physiol* **61**, 727–46
Goffredi M and Liveri V T (1981) *J Soln Chem* **10**, 695
Good W A (1967) *Nature* **214**, 1250–2
Hansch C and Leo A (1979) *Substituent Constants for Correlation Analysis in Chemistry and Biology*. New York, Wiley
Johnson C D (1973) *The Hammett Equation*. Cambridge, Cambridge University Press
Katz Y and Diamond J M (1974a) *J Membr Biol* **17**, 87–100
Katz Y and Diamond J M (1974b) *J Membr Biol* **17**, 101–20
Kavanagh F (ed) (1963) *Analytical Microbiology*. New York, Academic Press
King E J (1965) *Acid Base Equilibria*. Oxford, Pergamon Press
Krug R R, Hunter W G and Grieger R A (1976a) *J Phys Chem* **80**, 2335–41
Krug R R, Hunter W G and Grieger R A (1976b) *J Phys Chem* **80**, 2341–51
Krug R R, Hunter W G and Grieger R A (1976c) *Nature* **261**, 566–7
Krug R R, Hunter W G and Grieger-Block R A (1977) *ACS Symp Ser* **52**, 192–218
Leo A, Hansch C and Elkins D (1971) *Chem Rev* **6**, 525–616
Lissi E A (1981) *Acta Sud Amer Quim* **1**, 69–73
Nogrady T (1985) *Medicinal Chemistry*. Oxford, Oxford University Press
Rytting J H, Davies S S and Higuchi T (1972) *J Pharm Sci* **61**, 816–8
Saket M M, James K C and Kellaway I W (1984) *Int J Pharm* **14**, 155–66
Smith H J and Williams H (1983) *Introduction to the Principles of Drug Design*. Bristol, Wright
Tanford C (1980) *The Hydrophobic Effect: Formation of Micelles and Biological Membranes*. New York, Wiley-Interscience
Taylor J B and Kennewall P D (1981) *Introductory Medicinal Chemistry*. Chichester, Ellis-Horwood
Tomlinson E (1983) *Int J Pharm* **13**, 115–44
Wells P R (1968) *Linear Free Energy Relationships*. London, Academic Press

Chapter 9

Application of Calorimetric Methods in Ecological Studies

J. Widdows

9.1 Introduction

A thermodynamic energetic approach is widely used in ecological research to study the functioning of biological systems in relation to their environment. Energy flow is an important ecological concept concerned with the intake, conversion and passage of energy through an organism or an ecosystem. Thermodynamics can be studied at all levels of biological organization, from the energy exchanges within the cell (bioenergetics), to the whole organism (physiological energetics) and energy transfers between populations (ecological energetics). There have been comprehensive reviews of the thermodynamic concepts in the field of bioenergetics (Lehninger 1965), physiological energetics (Kleiber 1961) and ecological energetics (Brody 1945; Slobodkin 1962; Phillipson 1966; Wiegert 1976).

A comprehensive review of thermal methods and their application in ecological research is beyond the scope of this chapter. This review is therefore limited to the application of microcalorimetry in aquatic ecological research and the contribution calorimetry has made to some recent advances in the field of physiological energetics of aquatic animals.

9.1.1 Energy: Relations and Measurements

Production of matter (growth and reproduction) is a fundamental property of all living organisms and one that is necessary if a population is to persist in a given environment. The amount of production represents the difference between an individual's or a population's intake and output of matter or energy, and this will vary under different environmental conditions.

All organisms must work to live and therefore they require a source of potential energy which can be utilized. All work, including activity, growth and reproduction of living organisms, represents the transformation of energy and ultimately involves heat production.

Following the laws of thermodynamics and assuming steady state conditions (Kleiber 1961), the energy exchanges in an animal may be described by the balanced energy equation of Winberg (1960):

$$C - F = A = R + U + P$$

where C is the energy equivalent of food intake, F the energy content of faecal material, A the energy absorbed from the ingested food, R the energy equivalent of metabolic heat loss, U the energy content of excreta and P the energy incorporated into somatic growth and reproductive products.

All the processes that comprise the energy balance equation, namely, rates of feeding, food absorption, metabolism, excretion and production, are capable of variation in response to changes in the environment. The animal must be able to balance its energy gains from the environment against its metabolic losses in order to allocate surplus energy to somatic growth and reproduction. The evaluation of metabolic energy loss is a major aim in all energetic studies and for the construction of energy budgets. Total metabolic energy expenditure may be subdivided into maintenance metabolism, the energy required for maintenance of body functions and structures, and the additional energy-demanding activities associated with processes such as movement, energy acquisition (feeding and digestion of food) and growth.

9.1.2 Calorimetry

Calorimetry is a non-specific method for the direct measurement of metabolic activity by means of determining the rate of heat dissipation of biological processes. It detects the enthalpy changes of all reactions, even if the nature of these reactions is not fully understood. In order to increase specificity and understanding of complex biological processes, it is necessary to combine direct calorimetry with simultaneous indirect calorimetry (e.g. chemical and biochemical measurements).

Metabolic rate is routinely measured by indirect calorimetric methods, such as respirometry (O_2 uptake or CO_2 production) and biochemistry (utilization of metabolic substrates or accumulation of metabolic end products). However, reliance on indirect calorimetry has important limitations. Respirometry relies on standard values of heat equivalents (e.g. an oxycaloric equivalent of $-455\,\mathrm{kJ\,mol^{-1}\,O_2}$ to convert rates of oxygen consumption to heat dissipation) and this is based on the assumption that metabolism is totally aerobic, in a steady state, and without any anoxic component. Quantification of metabolic rate from biochemical measurements may be difficult if short term responses involve relatively small changes that are not easily detected and if unknown metabolic pathways are operational. The main limitation of direct calorimetry is the slow throughput and measurement of samples, which prevents its use for routine measurement of metabolic rates.

Energetic studies routinely use another form of calorimetry, known as bomb calorimetry (Crisp 1971). This method determines the heat of combustion (energy content) of biological materials (fuels, foods and production products) and is used to estimate the energy gains and non-metabolic energy losses in an energy budget, and the study of energy transfers in ecological systems.

This chapter is mainly concerned with the application of direct and simultaneous indirect calorimetry in quantifying and interpreting metabolic energy expenditure of aquatic organisms in response to different physiological states and environmental conditions.

9.2 Microcalorimetric Methods

Recent developments of microcalorimetry and experimental procedures have been described by Suurkuusk and Wadsö (1982). Microcalorimetric measurements are commonly performed in a closed ampoule or a flow-through mode. The most versatile ecological applications of microcalorimetry are offered by heat-flow calorimeters with perfusion (open-flow) animal chambers. 'The heat-flow principle ensures practically isothermal conditions for experimental periods of unlimited length. Whereas continuous perfusion provides a means for maintaining or manipulating the organism's environment in investigations of metabolic responses to environmental variables' (Gnaiger 1983b).

Two open-flow respirometer-microcalorimeters employing LKB microcalorimeters (LKB 2107 and LKB 10700) with animal chambers of approximately $1{\cdot}5\,cm^3$ and perfused with flow rates between 3 and $30\,cm^3\,h^{-1}$ have been described by Gnaiger (1983a) and Knudsen *et al* (1983). The volume of the animal chamber and the maximum flow rate has limited their use to animals of biomass less than about 20 mg dry mass (*Tubifex*, *Mytilus*, *Cyclops* sp.).

A perfusion chamber has recently been developed by Widdows and Salkeld (unpublished) for use with the latest LKB microcalorimeter (BioActivity Monitor, BAM; Suurkuusk and Wadsö 1982) The design of the perfusion chamber for the $25\,cm^3$ measuring cylinder of the BAM is illustrated in Figure 9.1. It uses a larger animal chamber ($25\,cm^3$) accommodating animals up to about 800 mg dry mass and relatively high flow rates up to at least $120\,cm^3\,h^{-1}$ (baseline noise $\pm < 4\,\mu W$) thus avoiding a reduction in O_2 content below 80% of air saturation in the animal chamber.

The major problem of equilibrating the temperature of the inflowing water to that of the microcalorimeter is overcome by a two stage temperature equilibration process. The first stage, external to the calorimeter, consists of passing the water through a 1 mm (i.d.) stainless steel tube within a temperature controlled water jacket that equilibrates the water to within $\pm 0{\cdot}1\,°C$

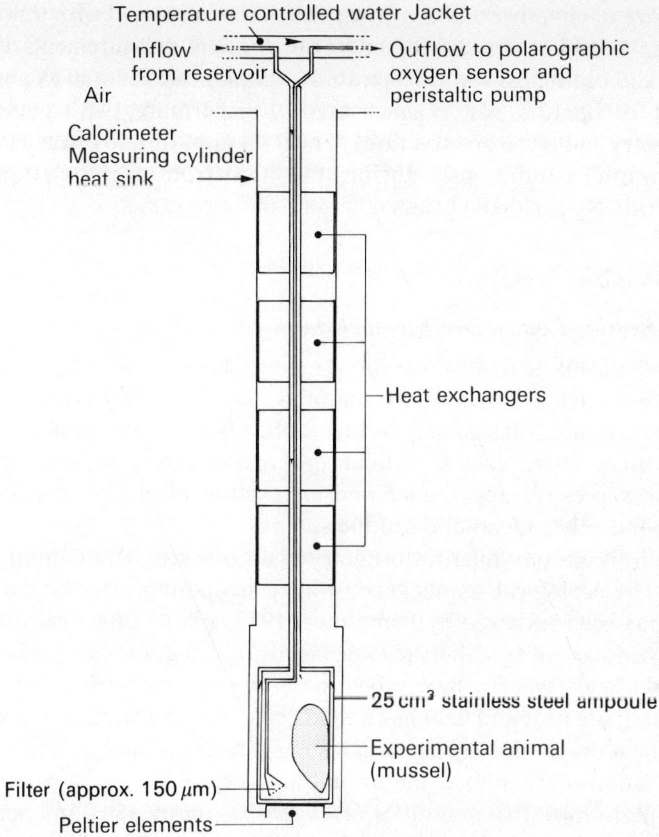

Figure 9.1. Diagram of 25 cm^3 perfusion chamber for use with a LKB microcalorimeter-BioActivity Monitor (for details see text).

of the operating temperature of the calorimeter. The second stage of temperature equilibration occurs as the water passes down a 3·1 mm (i.d.) wide stainless steel tube to the animal chamber (Figure 9.1). The increased residence time in the wider tube and the four heat exchangers, which provide close contact with the heat sink of the measuring cylinders of the BAM, equilibrates the water to the operating temperature of the calorimeter before it enters the animal chamber. The outflow is via a 1 mm (i.d.) stainless steel tube from the bottom of the chamber and up through the centre of the wider inflow tube. The outflow then passes away from the perfusion chamber and out of the BAM in a 1 mm tube, again carried in a temperature-controlled water jacket, to an oxygen sensor for simultaneous respirometric measurement. The system is calibrated by means of an internal calibration capsule which switches on and off a known power output at selected time intervals (e.g. 2 h).

The size of chamber and the flow rates that can be used with this perfusion chamber extends the scope of microcalorimetric measurements in physiological and ecological studies. Aerobic (aquatic) rates as well as anoxic rates of heat dissipation can be measured. Furthermore, with sumultaneous calorimetry and respirometry, rates of heat dissipation and oxygen uptake can be monitored continuously during transition from normoxia (aquatic) to anoxia (air/N_2 gas/deoxygenated water) and *vice versa*.

9.3 Microbial Ecology

9.3.1 Benthic Community Metabolism

Microorganisms generally occur in large numbers in soils and the sediments of aquatic systems. However, the majority may be either 'dormant or resting' and only a minor fraction may be metabolically active at any one time. There is a continual microbial succession as populations increase, peak and decline with the successive appearance and exhaustion of energy sources, growth factors and other metabolic products.

The application of microcalorimetry in marine sediment-community studies and the ecological significance of heat dissipation rates by natural sediments has been reviewed by Pamatmat (1977, 1982). Direct calorimetry has the advantage over other (biochemical) techniques for measuring the metabolic activity of complex benthic communities because all metabolic types dissipate heat and heat has a unidirectional flow without recycling. The rate of heat dissipation by one organism is directly comparable to that of any other organism. Therefore, the technique is especially useful with sediments that contain mixed communities of anaerobes, fermenters and aerobes, because the total metabolic activity of such communities can be measured. For example, heat dissipation rates of sediment from San Francisco Bay mud flat showed that approximately 50% of the total heat loss occurred in the surface sediments (0–1 cm), declining rapidly to zero at >6 cm depth (Pamatmat 1982). The surface sediments were exposed to dissolved oxygen and aerobic heat dissipation accounted for about 30% of the total heat dissipation by the surface layer (0–1 cm). While microcalorimetry offers high sensitivity, the technique provides a very non-specific measurement of heat loss (biological and chemical reactions) that can present problems in interpretation. This can be overcome to some extent, by manipulating experimental conditions and/or by using direct calorimetry in conjunction with other chemical measurements of metabolic rate.

Establishing relevant baselines to distinguish metabolic activity from possible heat flux associated with purely chemical reactions that might be taking place simultaneously (i.e. complexation and neutralization reactions) is often difficult (Ljungholm *et al* 1979; Pamatmat 1982). There are changes in chemical and physical properties of samples following sterilization by heat and ionizing radiation, and only partial inhibition by bactericides and fungicides.

The use of metabolic inhibitors commonly used in sediment oxygen-uptake measurements has not proved satisfactory in combination with direct calorimetry because of the long-lasting heat effects of unknown reactions between the poisons and reduced compounds and other substances in the sediment. Furthermore, the application of commonly used antibiotics will not delineate bacterial metabolism from metabolism of the rest of the benthic community (Yetka and Wiebe 1974).

Three methods of determining metabolic activity of marine sediments, namely, heat dissipation (Q) and two chemical measurements, adenosine triphosphate (ATP) concentration and the activity of the electron transport system (ETS), were compared by Pamatmat *et al* (1981). However, due to the problems associated with the measurement and interpretation of each of these responses there is no simple correlation between the techniques. Calorimetry measures metabolic energy flow in a complex sediment community in addition to heat associated with purely chemical reactions, whereas ATP concentration may provide an estimate of biomass and ETS is an *in vitro* measurement of respiratory enzyme activity and represents the maximum potential metabolic activity.

In spite of the problems concerning interpretation of complex power-time (*p-t*) curves of sediments and soils, *p-t* traces have been found to be quantitatively and qualitatively reproducible under specific environmental conditions, and microcalorimetry has been used to examine various aspects of metabolic activity associated with sediment and the water-sediment interface. Microcalorimetry has highlighted seasonal changes in the metabolic activity of complex mixtures of microorganisms colonizing the seawater-sediment (sandy-mud) interface in response to experimental eutrophication (Lasserre and Tournié 1984; Tournié and Lasserre 1984). The addition of protein-rich bactopeptone to the microcosm produced reproducible *p-t* traces which indicated marked seasonal differences between summer and winter (Figure 9.2). There were three phases: (*a*) the initial steady state at time of enrichment; (*b*) the exponential transitory phase after a variable lag time, which appears to be similar to classical batch cultures; and (*c*) the second steady state which varies seasonally. Lasserre and Tournié (1984) measured p_{O_2}, ATP levels and the incorporation of ^{14}C-glucose into $^{14}CO_2$ and ^{14}C-particulates, in an attempt to interpret the metabolic power-time curves. The release of $^{14}CO_2$ correlated with the *p-t* traces and reflected metabolic activity, whereas the ATP levels and ^{14}C-particulates were similar and reflected changes in biomass. The exponential phase of growth following nutrient addition was similar to classical microbial batch cultures with increases in all responses. However, after the transitory phase the system performed differently to batch cultures, in that the decline in heat dissipation was not correlated with changes in ATP, ^{14}C-particulates or p_{O_2}.

Direct calorimetry has also been used as a method of investigating the relative contribution of populations of component species to metabolic activ-

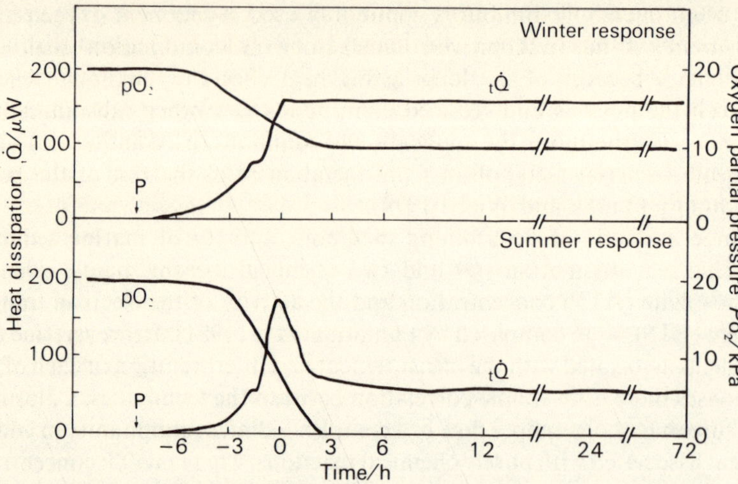

Figure 9.2. Typical examples of power-time ($_t\dot{Q}$) and oxygen-time (p_{O_2}) curves produced by an experimental eutrophication (P-peptone addition; 4 mg cm^{-3}) at the water-sediment interface: winter response (February–March; 12 °C); summer response (July–September; 22 °C). Response can be divided into three phases; (*a*) initial steady state; (*b*) a transitory phase after peptone treatment, a lag-time, an exponential phase and, except in winter, a decrease in heat dissipation; (*c*) a second steady state (after Lasserre and Tournié 1984).

ity and energy flow in benthic communities. A study by Pamatmat and Findlay (1983) demonstrated the potential role of heat flow measurements and experimental manipulations in understanding the biological interactions between species (microbes, nematodes and polychaetes) in a microcosm organically enriched with cereal. Species were found to have a significant impact on energy flow through a benthic community. For example, the presence of nematodes (*Diplolaimella chitwoodi*) depressed microbial metabolism and growth, whereas the polychaete (*Capitella capitata*) enhanced microbial metabolism and growth in organically enriched sediment, presumably by bioturbation, and consequently the rate at which microbial communities oxidize organic matter in sediment.

Microcosms represent semi-controlled systems that can be manipulated and perturbed to provide insight into the structure and dynamics of a benthic community. If a species has a significant impact on benthic energy flow, its addition to, or removal from that system will be reflected by a change in total heat dissipation. Direct calorimetry, therefore, provides a means of detecting a response of the entire system to various perturbations and manipulations. However, because of the generality of heat flux measurements, these alone cannot provide understanding of changes in the structure and function of benthic communities. Additional physical, chemical and biological parameters require measurement. Pamatmat and Findlay (1983) have emphasized the limitations of direct calorimetry in this field, but suggest that it may eventually be possible to formulate hypotheses about the ecological processes and the dynamics of benthic systems based on the nature of power-time curves, taking into consideration the frequency and amplitude of fluctuations and rates of change. Lasserre and Tournié (1984) have also suggested that a calorimetric approach may provide an opportunity to study temporal behaviour of complex microbial mixtures and the ability of populations to recover from induced perturbations such as excess organic matter, biocides or pollutants. However, the microbial-sediment studies employing direct calorimetry can only be regarded as preliminary attempts to investigate the complex dynamics of benthic systems, and energy flow associated with microbial succession. The work to date has been largely descriptive without a well-defined conceptual framework: therefore, the long-term potential of this approach is uncertain.

9.3.2 Pollution Studies and Toxicological Applications

There are very few examples of the application of microcalorimetry to problems of environmental pollution and toxicology. The work of Harju-Jeanty (1982) on the control of slime in paper-making waters provides a good example of the potential role of direct calorimetry in ecological/industrial management. Developments in the pulp and paper making industry with more closed circulation systems, larger paper mills and higher rates of production, have brought about an increased demand for a rapid and effective method of assessing and controlling microorganisms with antislime agents. Both biological and chemical slimes create major problems in the pulp and paper industries (e.g. blockages, corrosion, production stoppages and reduction in paper quality) and manufacturing becomes less profitable. The main causes of these production problems are the microorganisms (bacteria, moulds, yeasts and algae) in the process waters. Microcalorimetry has played an important role in establishing the minimum amounts of slimicide required to keep the microorganisms below a critical concentration, thus maximizing environmental protection, reducing the risk of producing slimicide resistant strains and minimizing the economic costs of slime control. Harju-Jeanty (1982) reviews the advantages and disadvantages of the different methods of studying slime problems in paper-making waters (nutrient agar, respirometric, photometric, bioluminescence and microcalorimetric). The microcalorimetric technique represents a significant improvement over the widely used nutrient-agar method, by providing a rapid measurement of total microbial activity in about 4 h under simulated paper-mill conditions, compared with 2 days for bacteria and 14 days for fungi using agar designed to meet specific nutrient requirements. Power-time traces were used to establish the inhibitory concentration and lethal dose for various slimicides on circulation water from Finnish paper-mills. Figure 9.3 illustrates the effect of increasing the concentration of the slimicide (48% *bis*-bromo-acetoxybutene) on the *p*-*t* trace. A

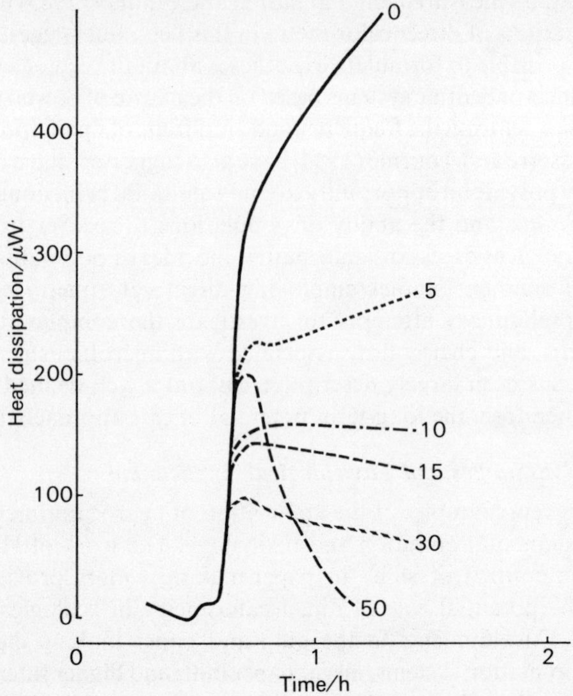

Figure 9.3. Effect of increasing concentration of slimicide (48% *bis*-bromo-acetoxy-butene) on power-time curves for the circulation water from a Finnish paper-mill. Added nutrient: 14 cm^3 nutrient broth solution (8 g NB/1000 cm^3 H_2O) in 1000 cm^3 circulation water. Microorganisms were incubated with slimicide for 0·5 h before measurement of heat dissipation (after Harju-Jeanty 1982). (Concentrations of slimicide in mg dm^{-3} given on curves).

curve shifting to a horizontal level indicates an inhibition of microbial growth and occurs at a concentration suitable for controlling microorganisms in the environment, whereas a curve declining to zero represents a lethal dose that in normal slime control is an unnecessarily high amount of slimicide. The inhibitory concentration, established by the microcalorimetric method, has been found to be empirically optimal and in the order of 1–25 ppm, depending on specific slimicide, or 15–36-fold lower than the inhibitory concentration established by the nutrient agar method. The ability to differentiate between bacteriostatic and bactericidal effects is most important from both economic and ecological considerations and with the aid of microcalorimetry it has proved possible to derive a realistic concentration of slimicide for controlling microorganisms below a critical concentration.

Microcalorimetry has also been used to determine the degradability of different substrates in different parts of a wastewater treatment system (Redl and Tiefenbrunner 1981). Total microbial metabolic activity was measured in terms of heat dissipation and this showed that the highest degradation rates occurred in the supernatant of the primary sludge and the degradation rates were progressively lower in the anaerobic digesting sludge and in the digested sludge from the storage area. It was concluded that these calorimetric measurements provided a non-specific, quantitative and continuous estimate of degradation processes that reflected how well adapted or poorly adapted the system was to the addition of new substrates. It was also considered to be better than measurements of either microbial growth or enzyme activity.

The use of flow microcalorimetry to monitor heat dissipation rates of monocultures of microorganisms may provide a sensitive bioassay technique for testing the effects of toxicants on their metabolic activity and growth. Flow microcalorimetry may also be a suitable technique for investigating growth control mechanisms such as those involving the balance between stimulatory (hormesis) and inhibitory responses of organisms (Stebbing 1982).

9.4 Aerobic and Anoxic Metabolism in Aquatic Invertebrates and Fish

Many aquatic invertebrates and some vertebrates possess metabolic adaptations enabling them to tolerate extended periods of anoxia and to exploit oxygen-depleted environments. Anoxibiosis can be divided into two basic forms, namely, 'environmental' due to oxygen depletion in the animal's immediate surroundings and 'physiological or functional' resulting from intense muscular activity.

Environmental anoxia in experienced by many sessile intertidal marine invertebrates, where they are subjected to regular periods of exposure to air, a medium in which most are morphologically ill-equipped to respire without problems of desiccation. Environmental anoxia is also experienced by benthic animals living in freshwater and marine systems that are subject to eutrophication, with high organic inputs and 'biological oxygen demand' resulting in hypoxia/anoxia in the water column and surface sediments. The biochemical responses to environmental hypoxia/anoxia have recently attracted considerable scientific interest (Hochachka and Somero 1984) and have been studied particularly thoroughly in the bivalve mollusc, *Mytilus edulis*, (de Zwaan 1977, 1983) and in parasitic helminths (Saz 1981). Biochemical adaptation to environmental anoxia is characterized by extremely low levels of metabolism combined with high efficiency of anoxic energy conversion.

In contrast, physiological anoxia occurs during periods of intense muscular activity, when aerobic metabolism cannot meet the increased energy requirement. 'Rapid muscular contraction is an explosive process, powerful but exhaustive, usually in a matter of seconds to minutes' (Gnaiger 1983c). Essentially anoxic biochemical pathways are employed (Kushmerick 1977; di Prampero 1981), and physiological anoxia is characterized by high power output and low efficiency.

Direct calorimetry represents the most general approach to estimating total biological energy conversion under conditions ranging from normoxia to anoxia. A combination of calorimetric, respirometric and biochemical methods makes it possible not only to separate total metabolism into aerobic and anoxic components but also to provide a thermodynamic interpretation of heat.

Factors affecting respiration rates of aquatic animals have been extensively reviewed (Newell 1979; Brett and Groves 1979; Bayne and Newell 1983) and therefore will not be considered in this chapter.

9.4.1 Oxycaloric Equivalents

In ecological studies metabolic energy loss is normally estimated by respirometric methods (indirect calorimetry). The concept of indirect calorimetry assumes that directly measured rates of total heat dissipation (${}_{t}\dot{Q}$) correspond to rates that are calculated on the basis of metabolic changes (${}_{m}\dot{Q}$). Oxycaloric equivalents ($\Delta_k H_{O_2}$) are used to convert rates of oxygen consumption ($\dot{N}_{O_2}$; i.e. molar amount of O_2 per unit time) into rates of catabolic heat dissipation (${}_{k}\dot{Q}$), so that:

$$ {}_{k}\dot{Q} = \dot{N}_{O_2} \times \Delta_k H_{O_2} \tag{9.1} $$

The oxycaloric equivalent in aerobic catabolism, $\Delta_k H_{O_2}$, is the energy per mole O_2 consumed and ranges from about -440 to -480 kJ mol^{-1} O_2 in aquatic animals (Gnaiger 1983c). However, the generalized oxycaloric equivalent ($\Delta_k H_{O_2} = -450$ kJ mol^{-1} O_2), as commonly used to calculate the metabolic energy loss in respiratory studies and for the construction of energy budgets, relies on a number of important assumptions. Metabolism is assumed to be in steady state with the dissipative (catabolic) conversion of substrates to CO_2, water and nitrogenous end products, and with intermediates remaining constant. Under normal physiological conditions the rate of ATP generation is balanced by ATP consumption and there are no net changes in the concentrations of high energy intermediates (ATP, phosphoarginine, etc). Dissipative catabolism prevails under various physiological conditions, when endergonic and exergonic half-cycle reactions keep high energy intermediates at steady-state concentrations and anabolic half-cycle reactions (*a*) are complemented by the dissipative half-cycle (*d*) (Figure 9.4; Gnaiger 1980b; Gnaiger *et al* 1986). This is the case in protein turnover, muscle contraction and relaxation, futile biochemical cycles, etc. Net biosynthesis, however, gives rise to (endothermic) anabolic heat changes (${}_{a}\dot{Q}$) characterizing conservative metabolism (Meyerhof 1920; Gnaiger 1983c).

Consequently, there are physiological and environmental conditions when the generalized oxycaloric equivalent is inappropriate, and there are apparent discrepancies between the expected oxycaloric equivalent and the experimentally determined caloric quotient of oxygen consumption.

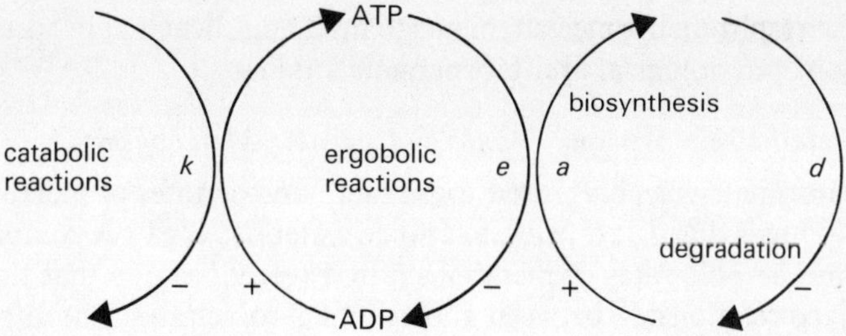

Figure 9.4. Coupling of metabolic half-cycle reactions in dissipative and conservative metabolism (*k*—catabolic; *e*—ergobolic; *a*—anabolic; *d*—degradative). Total heat dissipation is the sum of all net enthalpy changes (− exothermic; + endothermic) (after Gnaiger 1980b and Gnaiger *et al* 1986).

If the experimentally derived oxycaloric equivalent exceeds the range of theoretical $\Delta_k H_{O_2}$ values (Gnaiger 1983c) then partial anoxibiosis is indicated. The higher the experimental oxycaloric value the greater the reliance on anoxic metabolism. For example, a total heat equivalent of $-1250\ \mathrm{kJ\ mol^{-1}\ O_2}$ was recorded for air-exposed intertidal mussels, *Mytilus edulis*, (Widdows and Shick 1985). On the other hand, if the total heat equivalent of oxygen falls below the theoretical range of $\Delta_k H_{O_2}$ values then net biosynthesis giving rise to endothermic anabolic heat changes are suggested. Low values of $-252\ \mathrm{kJ\ mol^{-1}\ O_2}$ are apparent during the early stages of recovery from anoxia when there is an elevated rate of oxygen uptake, termed the oxygen debt payment (Shick and Widdows 1981; Gnaiger *et al* 1986).

The total heat equivalent of oxygen can also be altered by toxicants and a marked increase has been interpreted erroneously as an uncoupling of the respiratory chain. Gnaiger (1981) emphasizes that uncoupling *per se* cannot increase the oxycaloric equivalent but increases the caloric equivalent of ATP turnover by decreasing the $\sim P/O_2$ ratio (phosphorylation quotient). Therefore the same amount of heat is dissipated for the same amount of substrate and oxygen metabolized, but less ATP is generated. Consequently, an increase in calorimetric and respirometric rates does not indicate an equal increase of metabolism in terms of ATP turnover, if uncoupling occurs.

In spite of these recorded differences in total heat equivalent of oxygen, experiments employing simultaneous direct calorimetry and respirometry have demonstrated that the heat equivalent to oxygen consumption corresponds to approximately $-450\ \mathrm{kJ\ mol^{-1}\ O_2}$ for aquatic animals including bivalves, gastropods, crustaceans and fish under steady state normoxic conditions (Famme *et al* 1981; Gnaiger 1983a; Pamatmat 1983b). This confirms that metabolism is predominantly aerobic (>95%) with no significant contribution from anoxibiosis and that the standard oxycaloric equivalent used to convert respirometric measurements to metabolic heat is appropriate in the majority of physiological and bioenergetic studies.

9.4.2 Relationship between Anoxic and Aerobic Metabolism

Direct calorimetry has been used to measure anoxic rates of metabolism by aquatic animals during exposure to anoxic water or N_2 gas. A comparison of anoxic and aerobic rates of metabolism in Table 9.1 shows that there is no simple fixed relationship between anoxic and aerobic rates. The anoxic rates relative to the rate of heat dissipation ranges from 4% for *Mytilus edulis* (Widdows and Salkeld 1984) to 97% for *Mulinia lateralis* (Shumway *et al* 1983).

The sessile marine mussel (*Mytilus edulis*) living in the intertidal environment creates an anoxic/hypoxic microenvironment within its closed shell valves during air exposure. It therefore becomes inactive and the rate of energy expenditure is reduced to a very low anoxic rate. Periodic reliance of anoxibiosis as an energy saving mechanism has been suggested earlier by de Zwaan (1977) and Pamatmat (1980) and subsequently demonstrated in terms of an energy budget by Widdows and Shick (1985). Sessile and infaunal bivalves generally show a strong resistance to anoxia due in part to a reduction in activity and hence energy utilization.

In contrast, the clam *Mulinia lateralis* is a mobile, opportunistic species with a short generation time, adapted to life in 'transient' anoxic environments characterized by unstable, reducing substrate and organic pollution. Unlike most bivalves, *Mulinia* maintains high levels of feeding, shell valve and locomotory activities under anoxia, which may be an adaptation to escape periodic burial in unstable, oxygen-deficient sediment. This implies a high anoxic energy demand which is confirmed by direct calorimetry (Shumway *et al* 1983). Although the rate of metabolic heat dissipation under anoxia is the same as under normoxia, because of the much lower 'biochemical efficiency' or ATP yield per glycosyl unit in anoxic metabolism (4·71 mol ATP per mol of glycogen; succinate pathway) compared to aerobic metabolism (37 mol ATP per mol of glycogen; Gnaiger 1983c), glycolytic flux must increase (Pasteur effect) in order to maintain levels of ATP production during anoxic exposure. Moreover, due to the lower caloric equivalent of dissipative ATP turnover in anoxic metabolism ($-43\,\text{kJ}\,\text{mol}^{-1}$ ATP) compared to aerobic metabolism ($-80\,\text{kJ}\,\text{mol}^{-1}$ ATP; Gnaiger 1980b, 1983c) an equal rate of heat dissipation under both conditions implies a greater rate of ATP turnover under anoxia and therefore an 8-fold increase in glycolysis in order to maintain anoxic heat dissipation (i.e. $-2879\,\text{kJ}\,\text{mol}^{-1}$ of glycosyl unit catabolized aerobically compared to $-331\,\text{kJ}\,\text{mol}^{-1}$ glycosyl unit catabolized anaerobically to succinate; Gnaiger 1983c).

Calorimetric studies, therefore, provide examples of two contrasting forms of metabolic adaptation to environmental anoxia in bivalves. The extreme flexibility in metabolic rate shown by *Mytilus edulis* enables energy

Table 9.1 Comparison of steady state anoxic and aerobic rates of heat dissipation in aquatic animals

Group	*Species*	*Rate of heat dissipation mW g^{-1} dry mass*		*Anoxic as % of aerobic*	*Temp./°C*	*Reference*
		Anoxic	*Aerobic (aquatic)*			
Coelenterates	*Actinia equina* (intertidal anemone)	0·083[b]	1·167	7%	15	Shick 1981
Bivalves	*Mytilus edulis* (intertidal mussel)	0·072[a]	1·353	4%	15	Widdows and Salkeld 1984
	Mytilus edulis	0·105[b]	0·921	11%	20	Famme *et al* 1981
	Modiolus demissus (intertidal mussel)	0·159[c]	2·12	7%	20	Pamatmat 1979
	Cardium edule (intertidal cockle)	0·29[c]	0·71	41%	20	Pamatmat 1980
	Arctica islandica (subtidal clam)	0·115[b]	0·492	23%	15	Pamatmat 1980
	Mulinia lateralis (subtidal clam)	3·24[b]	3·35	97%	15	Shumway *et al* 1983
Annelids	*Lumbriculus variegatus* (freshwater oligochaete)	0·19[b]	0.44	43%	12	Gnaiger 1980a
	Diopatra cuprea (polychaete)	0·538[b]	1·60	35%	20·5	Hammen 1983
Crustaceans	*Cyclops abyssorum*	1·0[b]	6·0	17%	6	Gnaiger 1981
Fish	*Salvelinus alpinus* (fish eggs)	0·25[b]	1·5	17%	8·5	Gnaiger 1979
Reptiles	*Pseudemys scripta*	0·3[b]	2·6	12%	24	Jackson 1968

[a] N_2 gas.
[b] N_2 purged or deoxygenated water.
[c] Sealed animal in air.

expenditure to be reduced to low anoxic rates of metabolism, thus conserving energy and allowing the species to tolerate prolonged anoxia (e.g. about 35 days; Theede *et al* 1969). In contrast, the high anoxic rate of metabolism by *Mulinia lateralis* results in a high 'cost' and rapid depletion of glycogen reserves, which presumably accounts for the species' relative intolerance of prolonged anoxia (i.e. about 5 days; Shumway *et al* 1983).

9.4.3 Metabolic Rate–Body Size Relationship

The relationship between aerobic metabolism and the body mass of invertebrates and vertebrates is well documented (reviewed by Prosser 1973) and generally follows the 'surface-law' theory with metabolic rate increasing with 2/3 to 3/4 power of body mass. Therefore, respiration appears to be proportional to body or cell surface area available for O_2 exchange, rather than body mass.

In contrast, there is relatively little information in the literature concerning

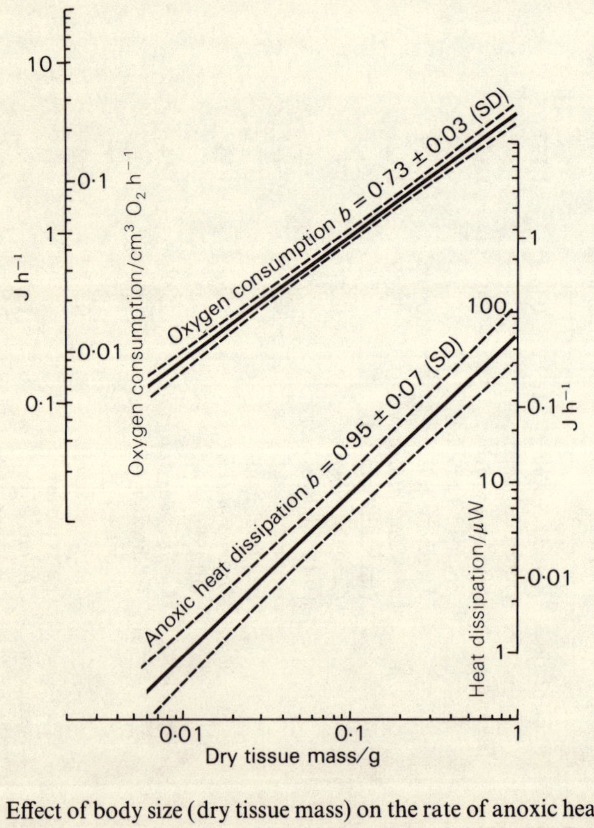

Figure 9.5. Effect of body size (dry tissue mass) on the rate of anoxic heat dissipation in N_2 gas and aquatic respiration by *Mytilus edulis* (March 1983; 12 °C; mean ± 95% C.L.).

the relationship between anoxic metabolism and body mass. Pamatmat (1979) pooled rather limited data ($n = 11$) on the rate of anoxic heat dissipation by two species of bivalves (*Polymesoda caroliniana* and *Modiolus demissus*) and concluded that anoxic metabolism was proportional to body mass. In a more detailed study, the relationship between anoxic heat dissipation and body size of *Mytilus edulis* was found to have a slope not significantly different from unity, indicating that the anoxic rate was indeed proportional to body mass (Figure 9.5; Widdows and Salkeld 1984). Clearly the relationship between anoxic metabolism against body mass is fundamentally different from that for aerobic metabolism against body mass in which the '*b*' value, the power coefficient of body mass, tends to a value of approximately 0·7, indicating that respiration is limited by the 'surface area' available for O_2 diffusion. Anoxic metabolism does not suffer from such constraints and thus has a rate proportional to body mass.

The presentation of aerobic metabolism as a mass-specific rate ($\mu W g^{-1}$), however, shows that the 'text-book' relationship of decreasing mass-specific rate with increasing body mass does not extend back to the gametes and early larval stages. Microcalorimetric measurements of aerobic rates of heat dissipation by gametes and larval stages of *Mytilus edulis* (Sprung and Widdows 1986) show that there are two components to the curve describing the relationship between mass-specific heat dissipation and body mass (Figure 9.6). The left-hand component represents an increase in mass-specific metabolic rate with early larval development from gametes to zygote through to the veliconcha stage. The later developmental stages follow the typical metabolicbody size relationship described by earlier workers (Zeuthen 1947;

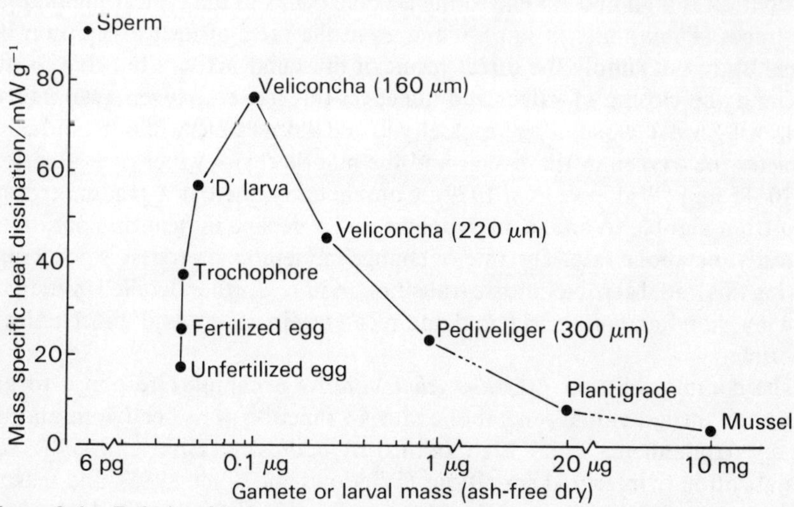

Figure 9.6. Relationship between mass-specific rate of heat dissipation (normoxia, 15 °C) and biomass (ash-free dry mass) during development of *Mytilus edulis* from gametes through larval stages to small immature mussels.

Hemmingsen 1960) when ${}_t\dot{Q}$ declines with increasing body mass according to an exponent value of $-0{\cdot}23$ (i.e. $1-b$). The maximum mass-specific rate appears to coincide with the larval stage which is most actively swimming, feeding and growing. Measurements of metabolic rates of bivalve larvae by direct calorimetry provide consistently higher estimates (about 2-fold) compared with rates in the literature based on indirect techniques such as Cartesian divers and oxygen sensors. The difference is probably the result of abnormal behaviour and feeding at the higher densities used in the small volume chambers associated with the indirect techniques.

9.4.4 Behavioural Effects

Power-time curves show major fluctuations in amplitude, periodicity and length of active periods, which reflect changes in the behaviour and activity of the animal under both anoxic and normoxic conditions. Behaviourally induced changes in the metabolic rates of marine animals have been described by Pamatmat (1979, 1980, 1982, 1983a,b). Bivalves undergo irregular and sporadic partial opening of the valves in order to 'test' the environment for the presence or absence of water, oxygen, food, toxicants, etc.

Simultaneous measurement of rates of heat dissipation and oxygen uptake show that major fluctuations in *p-t* traces between anoxic and aerobic rates of heat dissipation coincide with the opening and closing of shell valves of bivalves such as *Arctica islandica*, *Modiolus demissus* and *Mytilus edulis* (Pamatmat 1980, 1983a; Shick *et al* 1986). Individual mussels with shell valves wedged open, or cut permanent openings (about 1 cm^2) in the valve, not only show an absence of fluctuations in the *p-t* traces but their rate of heat dissipation is high and similar to the aerobic peaks in the typical fluctuating *p-t* traces (Pamatmat 1983a,b). Changes in the rates of heat dissipation by mussels are not simply the direct result of muscular activity but due to the opening and closing of valves and the effect this has on oxygen availability. Following valve closure, the mussel will remain aerobic while it gradually depletes the oxygen in the tissues and the mantle cavity water over a period of 10–15 min (Widdows *et al* 1979). Consequently, there is a gradual transition from aerobic to anoxic metabolism and a decline in heat dissipation to a steady metabolic rate. The rate of change in metabolic processes occurring during this initial aerobic-anoxic transition requires further detailed investigation by simultaneous direct and indirect (respirometry and biochemical) calorimetry.

The common mussel (*Mytilus edulis*) shows a complex response to air exposure; the individual's metabolic rate is a function of its behaviour and/or energy requirements which are modified by body size (Gnaiger 1983b), acclimatization to intertidal conditions (Widdows and Shick 1985), and season (Widdows and Salkeld unpublished data). The majority of individuals close their valves when exposed to air, but some individuals voluntarily break their isolation by gaping, which is accompanied by a sharp increase in the *p-t*

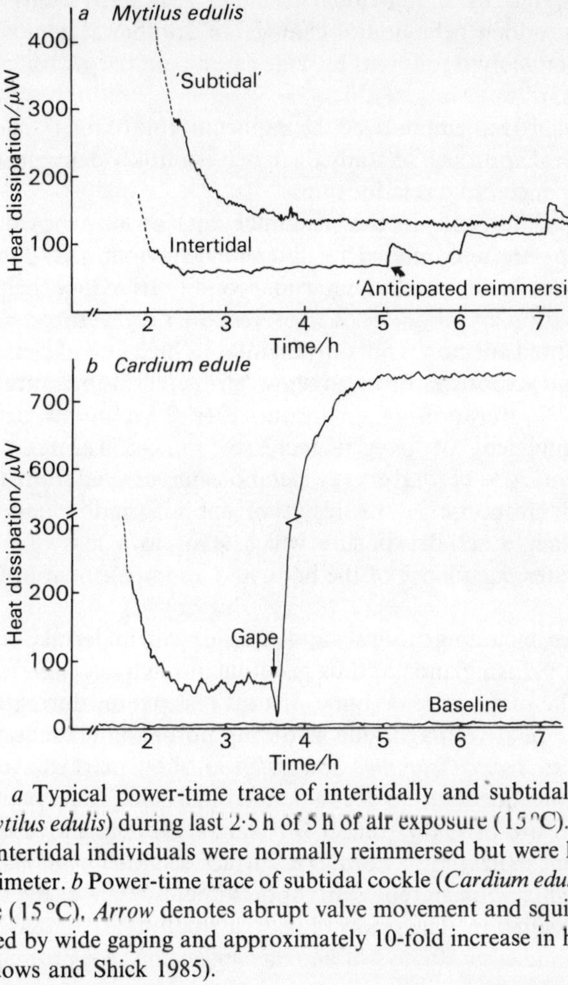

Figure 9.7. *a* Typical power-time trace of intertidally and subtidally acclimatized mussels (*Mytilus edulis*) during last 2·5 h of 5 h of air exposure (15 °C). *Arrow* denotes time when intertidal individuals were normally reimmersed but were left air-exposed within calorimeter. *b* Power-time trace of subtidal cockle (*Cardium edule*) during acute air exposure (15 °C). *Arrow* denotes abrupt valve movement and squirting of mantle fluid followed by wide gaping and approximately 10-fold increase in heat dissipation (from Widdows and Shick 1985).

trace (Gnaiger 1983b). This gaping behaviour is particularly pronounced in the smaller size-class (1–2 cm shell length; about 0·02 g dry tissue mass) compared to the larger individuals (4 cm; about 0·5 g dry tissue mass). Intertidally acclimatized mussels also show distinct activity peaks at the 'anticipated' time of reimmersion (Figure 9.7a; Widdows and Shick 1985) when specimens are held in the calorimeter beyond the normal reimmersion time. The aerial rate of heat dissipation is typically approximately 15% of the aquatic aerobic rate in larger individuals.

In contrast, the cockle (*Cardium edule*) typically gapes during aerial exposure and has a total heat equivalent of oxygen of $-430\ \mathrm{kJ\ mol^{-1}\ O_2}$, indicating that metabolism is totally aerobic (Widdows and Shick 1985). However, an atypical response by an individual cockle (Figure 9.7b) seems to illustrate the effect of a sudden behavioural change, an abrupt valve movement and squirting of mantle fluid followed by wide gaping, on the *p-t* trace (Widdows and Shick 1985).

Gnaiger (1983b) has emphasized the important role of microcalorimetry as the only practical approach to study these behaviourally determined complex anoxic/aerobic metabolic relationships.

Intraspecific differences in the metabolic rate of air-exposed anemones (*Actinia equina*) are also related to different behavioural responses (Shick 1981). Anemones acclimatized to intertidal conditions reduce their aerial rate of heat dissipation by about 30%, but remain fully aerobic. Conversely, subtidally-adapted anemones not only maintain a high rate of heat dissipation during acute air exposure, but also show an increasing reliance on anoxic metabolism with duration of emersion; after 2·5 h of air exposure the total heat equivalent of oxygen increases to $-550\,\mathrm{kJ\,mol^{-1}\,O_2}$, indicating that about 20% of total energy metabolism is derived from anoxibiosis. This metabolic response is the result of subtidal individuals remaining active during acute aerial exposure which leads to a loss of most of the coelenteron water, a collapse of the body and an impairment of O_2 delivery to the tissues.

Invertebrates, including molluscs and annelids, can undertake considerable physical activity during anoxia, thus resulting in relatively high rates of heat dissipation. The magnitude or burst of heat dissipation during anoxia can be of the same relative magnitude as during normoxia (Pamatmat 1983b). Bivalves such as *Arctica islandica* and *Solen* sp. show marked cyclic fluctuations in anoxic heat dissipation over a period of 10 days. These periods of high energy expenditure probably reflect the clam's effort to move through the anoxic sediment in search of normoxic surface sediments (Pamatmat 1980, 1983b). Active metabolism represents approximately 30% of the total rate of anoxic heat dissipation. The peaks of heat dissipation associated with bursts of activity decline after 10 days of anoxia, suggesting that prolonged anoxia has a limiting effect on the animal's capacity for bursts of activity. Even when exposed to normoxic conditions, *Arctica islandica* has an irregular behaviour pattern in which individuals may remain buried beneath the surface sediments for periods of 1–7 days: a spontaneous burial period as long as 24 days has been recorded (Taylor 1976). There is no apparent reason for this self-induced anoxia.

All benthic infaunal species in which metabolic rate has been measured both in and out of sediment have shown greatly elevated average rates when removed from sediment. The burrowing polychaete *Neanthes virens* shows large random and apparently abnormal fluctuations when held in water only, compared to a rhythmic activity in sediment (Pamatmat 1982). It is concluded that the avoidance of the complexity introduced by the presence of other living organisms in the sediment creates a greater error than performing two rate measurements of metabolism, one of sediment only and the other of experimental animal plus sediment (Pamatmat 1983b).

9.4.5 Temperature

Intertidal animals experience a wide range of temperatures and the influence of this environmental factor on the rate of oxygen consumption by marine animals has received considerable attention (reviewed by Newell 1979; Newell and Branch 1980). However, until the application of microcalorimetry for measuring total heat dissipation it was not possible to determine the effect of temperature on either the total energy expenditure during aerial exposure/anoxia or the relative contribution from aerobic metabolism. Two species of intertidal bivalves (*Polymesoda caroliniana* and *Modiolus demissus*), capable of tolerating elevated air temperatures in warm temperate environments, were found to have a relatively temperature independent rate of anoxic heat dissipation with a Q_{10} of 1·56 between 20 and 30 °C (Pamatmat 1979). The intertidal mussel *Mytilus edulis* shows a more temperature-dependent aerial rate of heat dissipation with a Q_{10} of 2·0–2·7 between 6 and 25 °C. Furthermore, the rates of heat dissipation by *Mytilus* during exposure to air or N_2 gas were similar with a slightly higher rate in air (Figure 9.8; Widdows and Salkeld 1984), suggesting that although there may be some diffusion of oxygen

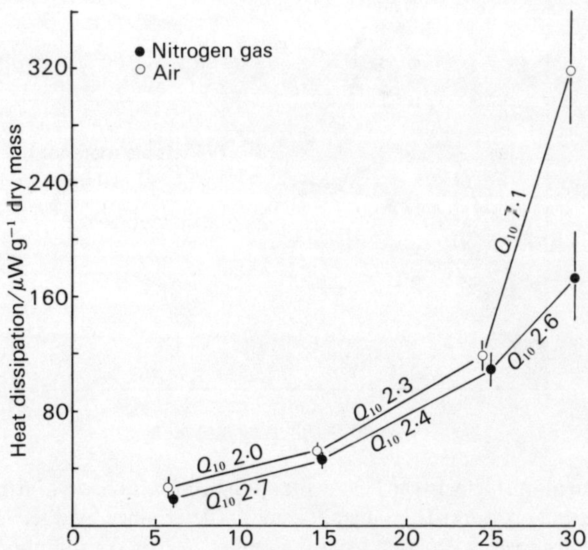

Figure 9.8. Effect of temperature on the rate of heat dissipation by *Mytilus edulis* (0.9 g dry tissue mass) during exposure (6–18 h) to air and N_2 gas (mean ± SE: $n = 8$).

between the closed shell valves the metabolic rate is predominantly anoxic. However, when mussels are exposed to high air temperatures of 30 °C (i.e. above the incipient lethal limit but within the zone of resistance) the aerial rate of heat dissipation is twice the anoxic rate due to the 'gaping' of the shell valves. This 'gaping' behaviour at elevated temperatures appears to be an adaptive response providing evaporative cooling in the natural environment (although not in the calorimeter). Consequently, the enhanced metabolic rate is probably a secondary effect and not a direct response to an intrinsically higher metabolic requirement, because mussels can survive both N_2 gas and air at 30 °C for at least 24 h. Such behaviour suggests that the need for cooling at elevated temperatures may outweigh the additional metabolic costs.

9.4.6 Hxpoxia

Many benthic animals are capable of physiological and biochemical adaptations to reduced oxygen availability in their environment. Physiological compensations to reduced oxygen tension include an increase in the percentage extraction of O_2 from the ventilation current and an increase in the blood

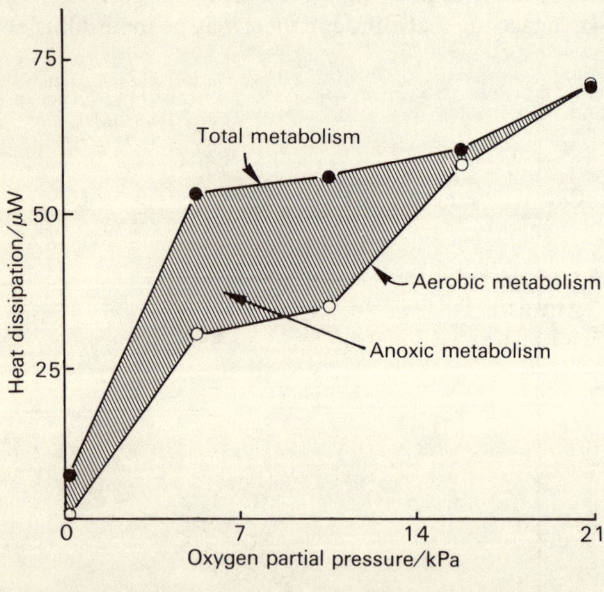

Figure 9.9. Relationship between heat dissipation by *Mytilus edulis* (7·6 mg dry mass) and oxygen partial pressure. Total metabolism was determined by direct calorimetry, aerobic metabolism was calculated from the rate of oxygen consumption multiplied by the oxycaloric equivalent ($-470\ kJ\ mol^{-1}\ O_2$), and the difference between total and aerobic metabolism represents anoxic metabolism (shaded area). Each point is the integrated rate of heat dissipation over about 20 h, measured at 20·25 °C and a salinity of 18·4% (after Famme *et al* 1981).

perfusion through respiratory surfaces. Biochemical adaptation to hypoxia involves the utilization of anoxic metabolism to supplement aerobic metabolism when insufficient O_2 can be extracted from the environment to meet the total metabolic requirements of the animal.

Simultaneous direct and indirect calorimetry has been used to study the relative levels of aerobic and anoxic metabolism in the mussel (*Mytilus edulis*) over a range of O_2 concentrations from full saturation to zero (Famme *et al* 1981). Total heat dissipation increases with increasing oxygen concentration to a maximum at normoxia (21 kPa) which is 8–10-fold higher than the anoxic rate. The aerobic component increases from zero at anoxia to approximately 100% of the total heat dissipation at normoxia. The anoxic metabolic component, based on the difference between the direct and indirect calorimetric measurements, increases with a rise in p_{O_2} from 0 to 10 kPa and then declines with a further rise in p_{O_2} to full saturation (21 kPa). Consequently, the total metabolic rate is maintained relatively independent of declining p_{O_2} down to 25–50% of air saturation by increasing the anoxic component which reaches a maximum between 5·33 and 10·66 kPa (Figure 9.9). A similar reduction in the rate of anoxic heat dissipation by *Salvelinus alpinus* in response to a decline in p_{O_2} from 48% to 19% air saturation was reported by Gnaiger (1983a). A reduction in the anoxic as well as aerobic rate of heat dissipation by *Mytilus* and *Salvelinus* at O_2 concentrations below about 25–50% air saturation is probably associated with the cessation of feeding and locomotory activity and hence a reduction in dissipative metabolism under severe hypoxia.

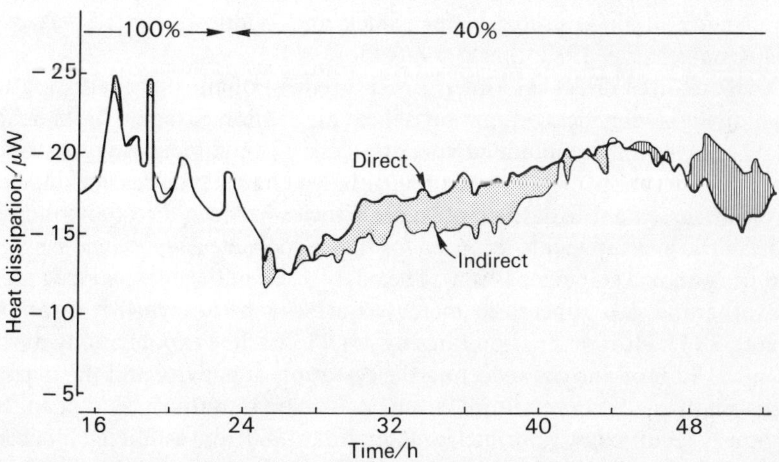

Figure 9.10. Simultaneous direct (thick line) and indirect (thin line) flow-calorimetric measurement of heat dissipation by one individual *Salvelinus alpinus* (5·11 mg dry mass; 8 °C, flow rate of 5·67 cm^3 h^{-1}) in response to a decline in p_{O_2} from 19·5 kPa to 9·3 kPa (= 100% to 48% of air saturation). There are two distinct phases in the acclimatory response to lowered p_{O_2}; (*a*) anoxic compensation (25–43 h; stippled area); and (*b*) conservative compensation (44–53 h; vertical shading) (after Gnaiger 1983a).

The use of direct and indirect calorimetry has also brought about the resolution of two distinct phases in the acclimatory response of *Salvelinus alpinus* to a reduction in p_{O_2} to 48% of air saturation (Figure 9.10; Gnaiger 1983a): (*a*) anoxic compensation (22–44 h) where aerobic sources of metabolic energy expenditure are augmented by anoxic processes (total heat equivalent ($\Delta_t Q_{O_2}$) = $-526\,\mathrm{kJ\,mol^{-1}\,O_2}$), thus increasing the ratio of measured to calculated heat dissipation (based on a generalized oxycaloric equivalent of $-450\,\mathrm{kJ\,mol^{-1}\,O_2}$), and (*b*) conservative (anabolic) compensation (48–52 h) where the ratio of measured and calculated heat dissipation falls below the aerobic value, indicating activation of coupled reactions in which part of the energy that would otherwise be expected to dissipate as heat is conserved ($\Delta_t Q_{O_2} = -388\,\mathrm{kJ\,mol^{-1}\,O_2}$). The explanation of these observations may be the fate and reutilization of accumulated anoxic end products (e.g. lactic acid) in aerobic respiration and coupled glyconeogenesis (Gnaiger 1983a). The apparent O_2 dependency of metabolic rate immediately after a decline in p_{O_2} to 48% air saturation is followed by a return to the normoxic level of metabolism within 1 day (Figure 9.10).

9.4.7 Thermodynamic and Biochemical Interpretation of Anoxibiosis and Recovery

Recent studies of the adaptive responses of the mussel (*Mytilus edulis*) to anoxia (up to 24 h exposure to air on N_2 gas) serve to illustrate the role of direct and indirect (respirometry and biochemistry) calorimetry in providing a metabolic interpretation of environmental anoxibiosis and recovery, i.e. reimmersion in air-saturated water (Shick and Widdows 1981; Shick *et al* 1983; Gnaiger *et al* 1987; Shick *et al* 1986).

Comparison of direct calorimetric measurement of anoxic heat dissipation and indirect biochemical estimation of heat dissipation based on the stoichiometric analysis of accumulated end products of anoxibiosis suggest that a large fraction (about 50%) of the total enthalpy change remains unexplained. This is in agreement with the large discrepancies between direct and indirect calorimetric measurements recorded for the oligochaete *Lumbriculus variegatus* during anoxia (Gnaiger 1980a). The relative size of the unexplained anoxic or exothermic gap appears to increase during a period of 48 h of anoxia (Figure 9.11). Heat dissipation rates by *Mytilus* decline exponentially during the initial 12 h of anoxia, reflecting the cessation of activity and the marked reduction in energy expenditure, a time course quantitatively similar to that previously recorded by calorimetry (Pamatmat 1980) and estimated biochemically (Ebberink *et al* 1979; Zurburg and Ebberink 1981). During anoxia the majority of bivalves effectively isolate themselves from their external environment and accumulate end products of anoxic metabolism in their body tissues (de Zwaan 1977, 1983).

Alanine is an initial end product but does not increase after 12 h. Succinate, however, accumulates continuously during anoxia and propionate forms only

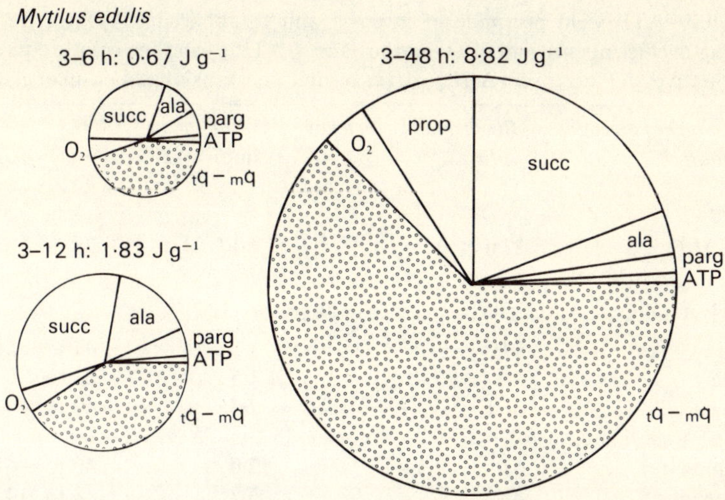

Figure 9.11. Anoxic heat dissipation in *Mytilus edulis* measured by simultaneous direct and indirect (biochemical) calorimetry. The area of each circle is proportional to the total heat change ($_t\dot{q}$/J g^{-1} dry mass) for that time interval (average rates are shown), and the identified sectors represent the metabolic heat change ($_m\dot{q}$/J g^{-1}) associated with various metabolic reactions. (ATP; parg—phosphoarginine; ala—alanine; succ.—succinate; prop—propionate; O_2—contribution to aerobic catabolism from depletion of physical and chemical O_2 stores). Approximately 50% of the measured heat dissipation is biochemically unexplained; stippled area represents the 'anoxic gap' ($_t\dot{q} - {_m\dot{q}}$) (after Gnaiger *et al* 1987).

after a lag period of at least 12 h. The depletion of ATP and phosphoarginine pools is relatively high at the onset of anoxibiosis and gradually declines to lower levels (Shick *et al* 1983; de Zwaan 1983). In long-term anoxia, the net change in high energy intermediates (ATP, phosphoarginine) become relatively unimportant (Shick *et al* 1983) and therefore metabolic heat dissipation is equivalent to dissipative catabolism.

The amounts of end products accumulated are used to calculate the ATP yield according to the stoichiometric equations of glycogen catabolism (Table 9.2). The relative efficiencies of the various anoxic glycolytic fermentations are extremely low, in terms of ATP yield per glucosyl unit, compared to complete degradation of carbohydrate by oxygen. Even in the case of propionate accumulation, which gives the highest ATP yield per glucosyl unit, this yield is reduced by a factor of 6. Therefore, both the route of glycogen fermentation and the rate of ATP turnover vary depending on the duration of anoxia. In addition, the caloric equivalent of dissipative ATP turnover under anoxia is not a constant but varies with particular pathways and side reactions involved (Gnaiger 1980b, 1983c). Anoxic heat dissipation is then estimated using the heat of dissipative ATP turnover associated with the overall reactions and the

Table 9.2 ATP yield per mole of glycosyl unit catabolized (^{v}ATP/S), the ATP coupling coefficients per mol of O_2 and product (^{v}ATP/p) and the caloric equivalent of dissipative ATP turnover ($\Delta_k H_{\infty ATP}$) in aerobic and anoxic catabolism of glycogen

End product of glycogen fermentation	*'Biochemical efficiency'* ${}^{v}ATP/S$	${}^{v}ATP/p$ /mol^{-1} O_2	${}^{a}\Delta_k H_{\infty ATP}$ /kJ mol^{-1} ∞ ATP
Aerobic			
$CO_2 + H_2O$	37·0	6·17	− 78
Anoxic			
lactate	3·0	1·5	− 47 to − 54
alanine	3·0	1·5	− 62 to − 69
succinate	4·71	2·75	− 49 to − 55
acetate (+succinate, propionate)	6·0	2·0	− 40 to − 45
propionate	6·43	3·75	− 38 to − 42

[a]Caloric equivalent of dissipative ATP turnover varies depending on the buffering systems and declines during anoxia due to accumulation of acids. From Gnaiger (1980b, 1983c).

heat changes of ergobolic reactions (ATP and phosphoarginine breakdown) (de Zwaan 1983; Shick *et al* 1983; Gnaiger *et al* 1987).

Shick *et al* (1983) suggested that the observed discrepancy between direct and indirect calorimetric measurements of anoxic heat dissipation is unlikely to be the result of any major energy-yielding anoxic metabolic pathways remaining undiscovered. However, there is recent evidence of ethanol, which together with NH_4^+, accumulates as an end product of anoxibiosis in some fish and a free-living nematode (Shoubridge and Hochachka 1980; Butterworth and Barnett 1985). The role of this and other anoxic metabolic pathways in aquatic invertebrates requires further study.

It is interesting to note that energy balance studies of vertebrate muscles have also shown that a large fraction of anoxic heat dissipation remains unaccounted for by the stoichiometric analysis of anoxic metabolism (Curtin and Woledge 1978).

The marked reduction in the rate of heat dissipation (about 4–7% of the aerobic rate) and the concomitant suppression of the Pasteur effect (de Zwaan and Wijsman 1976) represents an 'energy saving' mechanism during environmental anoxibiosis, but any evaluation of anoxic 'energy saving' must take into account the 'cost' of postanoxic recovery (Shick and Widdows 1981; Widdows and Shick 1985; Gnaiger *et al* 1987). Furthermore, any energy saving associated with reduced anoxic heat dissipation would be all in vain if glycolytic end products are then excreted or completely burnt during recovery and the oxygen debt payment.

The anoxic-aerobic transition or recovery from anoxia can be divided into two distinct phases; early recovery (0–1 h) characterized by low heat equivalents of oxygen (-366 kJ mol^{-1} O_2 or -252 kJ mol^{-1} O_2 when steady state ${}_t\dot{q}$ and $\dot{n}_{O_2}$ are subtracted) and late recovery (1–3 h) featuring high heat equivalents (-503 kJ mol^{-1} O_2) compared to the steady state value of -475 kJ mol^{-1} O_2. Immediately following reimmersion in air-saturated seawater the mussel opens its shell valves and there is a high rate of oxygen uptake which declines rapidly during the first hour and reaches a steady state value after about 3 h (Figure 9.12). This elevated rate of oxygen uptake in excess of the steady state value is referred to as the oxygen debt payment and is a common feature of recovery from environmental and physiological anoxia. Some of the excess oxygen uptake (approximately 15%) is involved in non-metabolic processes (Widdows and Shick 1985; Gnaiger *et al* 1987; Shick *et al* 1986) such as the physical reoxygenation of mantle cavity water, haemolymph and tissue water. An additional small component of the excess O_2 uptake may be associated with the chemical reoxygenation of carotenoid pigments, which have been suggested as an intracellular oxygen reserve in *Mytilus*. Even after correction for physical and chemical reoxygenation the total metabolic oxygen consumption during the first hour was twice that of

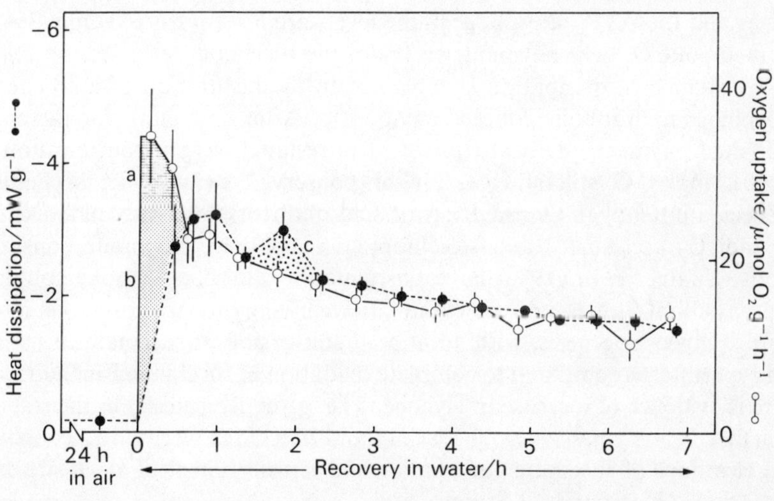

Figure 9.12. Mass-specific rates of heat dissipation (${}_t\dot{q}$/mW g^{-1}; solid circles) and oxygen uptake ($\dot{n}_{O_2}$/μmol O_2 g^{-1} h^{-1}; open circles) by *Mytilus edulis* during 24 h air exposure and aquatic recovery. The ${}_t\dot{q}$ and $\dot{n}_{O_2}$ scales are directly comparable assuming an oxycaloric equivalent of $-0{\cdot}125$ mW μmol^{-1} O_2 h^{-1}. Hatched area (a) represents portion of recovery oxygen uptake explained by reoxygenation of O_2 stores. Fine stippled area (b) represents period of excess O_2 uptake associated with endothermic recovery processes. Heavy stippled area (c) represents the period of anoxic metabolism during recovery (mean $\pm$ SE; $n = 4$ for ${}_t\dot{q}$; $n = 8$ for $\dot{n}_{O_2}$) (after Shick and Widdows 1981; Gnaiger *et al* 1987).

the steady state, whereas recovery heat dissipation was only 50% above the steady state. Consequently, there remains a large excess of net recovery oxygen consumption over recovery heat dissipation during the first hour of recovery (Figure 9.12), indicating conservation of heat in endothermic (anabolic) processes.

Further evidence of endothermic (anabolic) processes occurring during early recovery comes from biochemical studies (de Zwaan 1977; Widdows *et al* 1979) in which the size of the O_2 debt payment is considerably less than the amount of O_2 required for the complete oxidation of accumulated end products such as succinate, alanine and propionate. There is no evidence of excretion of succinate or alanine, but some propionate may be excreted and also reabsorbed by the mussel during recovery (Zurburg *et al* 1982).

Gnaiger *et al* (1987) have therefore attempted to explain these apparent discrepancies between direct and indirect (respirometric and biochemical) calorimetric measurements, by a stoichiometric and thermodynamic analysis of biochemical changes occurring in *Mytilus* during recovery from prolonged anoxia (de Zwaan *et al* 1983; Shick *et al* 1983). During the first hour of recovery, when about 60% of the O_2 debt payment is completed, approximately 50% of the succinate (the principal end product) is cleared from the tissues and the ATP, phosphoarginine levels are 50% restored. Only 7% of the metabolic O_2 debt payment, or 3% of the succinate clearance, or 5% of the recovery heat dissipation is explained in the restoration of high-energy phosphates in ergobolic coupled catabolism. After correction for the ergobolic debt payment, the heat equivalent of residual oxygen consumption of $-260\ \mathrm{kJ\ mol^{-1}}\ O_2$ still indicates a highly conservative (anabolic) mechanism of succinate removal. On the basis of acid quotients (mol succinate cleared per mol O_2 consumed) and stoichiometric and thermodynamic considerations, Gnaiger *et al* (1986) have postulated a catabolic-anabolic coupling whereby about 6 times more succinate can be discharged in catabolic-anabolic coupled glyconeogenesis, with additional conversion of succinate to malate and aspartate, as compared to complete oxidation of succinate. Furthermore, there is evidence of increase in glycogen (i.e. glyconeogenesis) in the starved clam (*Mercenaria mercenaria*) between 6 and 24 h of recovery with a concomitant clearance of succinate and alanine and replenishment of aspartate concentrations (Korycan and Storey 1983).

While most of the observed changes in biochemistry, heat dissipation and oxygen uptake can be explained, there still remains a biochemically unexplained oxygen debt payment with an endothermic component. The size of this 'endothermic gap' during recovery is in close agreement with the size of the 'exothermic gap' observed during anoxia (Gnaiger *et al* 1987), thus suggesting a compensatory reversal of the biochemically unexplained 'exothermic gap'. A similar reversal appears to occur during recovery of vertebrate skeletal muscle following exercise (Curtin and Woledge 1978; di Prampero

1981), with endothermic processes compensating for unknown exothermic processes which occur during anoxia.

In contrast to early recovery, during the later recovery period between 1·5 and 2·5 h there is a transient excess heat dissipation relative to oxygen consumption, reflected in a heat equivalent of $-503\,\text{kJ}\,\text{mol}^{-1}\,O_2$ or $-584\,\text{kJ}\,\text{mol}^{-1}\,O_2$ for the recovery heat equivalent when the steady state ${}_t\dot{Q}$ and $\dot{N}_{O_2}$ values are subtracted. This suggests the occurrence of anoxic metabolic processes, which precisely matches the time course for strombine, alanine and lactate accumulation in the posterior adductor muscle of *Mytilus* (de Zwaan *et al* 1983). Ongoing endothermic recovery reactions may mask additional anoxic (exothermic) heat changes which are then undetectable by the energy balance method. de Zwaan *et al* (1983) have suggested that the role of anoxic metabolism during recovery is to meet the increased energy demand and simply supplement aerobic energy metabolism which has limited scope during recovery. However, higher aerobic rates are attainable and oxygen is readily available and not diffusion-limited at this later stage of recovery. The function of this anoxic component of recovery metabolism may be the restoration of ATP and phosphoarginine pools, similar to the phenomenon of 'anaerobic recovery' in vertebrate skeletal muscle (di Prampero 1981).

This analysis of anoxia and recovery in *Mytilus edulis* (Gnaiger *et al* 1987) underlines the importance of combined thermodynamic and biochemical considerations in interpreting the rates of heat dissipation and the functioning of animals under aerobic and anoxic conditions.

9.4.8 Physiological Hypoxia

The energetics of muscular exercise in vertebrates has been reviewed by di Prampero (1981), but there have been few direct calorimetric measurements during physiological hypoxia in aquatic invertebrates, partly due to the difficulty in inducing intense activity inside calorimeters with chambers of limited volume. Physiological hypoxia cannot be discerned from purely aerobic mechanisms by direct calorimetry alone; the simultaneous measurement of oxygen consumption is required. In contrast to environmental anoxia, during physiological hypoxia/anoxia the rate of heat dissipation is high and typically in a non-steady state. Consequently, interpretation of power-time traces requires mathematical time correction when measuring rapid short-term changes, or integration of *p-t* traces over extended periods. Physiological hypoxia may be indicated by a correlation between high heat equivalents ($\text{kJ}\,\text{mol}^{-1}\,O_2$) and high total heat dissipation rates, suggesting a significant anoxic metabolic contribution during periods of spontaneous burst activity (Figure 9.13).

The energetics of physiological hypoxia/anoxia and recovery by aquatic animals is clearly a research area that would benefit considerably from the use of open-flow respirometry-microcalorimetry, to complement and advance the

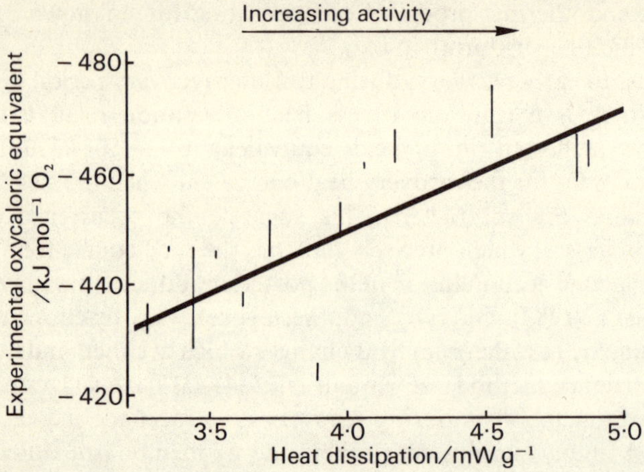

Figure 9.13. Simultaneous calorimetric and respirometric measurement of *Salvelinus alpinus* during spontaneous fluctuations of activity under normoxic conditions. Activity measured in terms of increased heat dissipation resulted in an increase in the experimental oxycaloric equivalent. The bars indicate the ranges of instantaneous values of the experimental oxycaloric equivalent due to activity dependent variations of the respirometer time constant (after Gnaiger 1983c).

interpretation of biochemical studies of energy metabolism in muscle tissues (reviewed by Livingstone 1982; Gäde 1983).

9.4.9 Anoxibiosis: Economy versus Power

Gnaiger (1983c) has examined anoxic metabolism in terms of energetic efficiency and economical power output and has thus extended the functional interpretation of anoxibiosis under physiological and environmental anoxia. The caloric efficiency of catabolic pathways increases sequentially in the course of aerobic-anoxic transition. Therefore, compared to aerobic metabolism, anoxic metabolism has a higher ATP gain per unit of heat dissipation which is equivalent to a lower catabolic heat dissipation at a constant rate of ATP consumption. However, the pathways with the lowest biochemical efficiency (ATP yield per mol of glycosyl unit catabolized) are the lactate, octopine, strombine and alanopine pathways (ATP mol^{-1} glycosyl = 3), which predominate during bursts of activity when high power output is required (physiological anoxia). In contrast, the succinate-propionate-acetate pathways, utilized under long-term environmental anoxia, exhibit the highest ATP yield (ATP mol^{-1} glycosyl = 4·7–6·43) because economy is more important than power.

The regulatory mechanisms controlling the conflict between economy and power in long-term anoxia have been discussed by Gnaiger (1983c).

Optimization involves a complex compromise: 1) the output rate, the rate of ATP production; and 2) the output force, the phosphorylation potential, must ensure a sufficient output power or ergobolic dissipation function to meet the maintenance requirements; 3) the thermodynamic efficiency has to be kept high if the anoxic endurance strategy aims at a high reversibility of metabolic processes for the restoration of the glycogen pool during the aerobic recovery period; 4) the most effective use of the substrate store with minimum load on the end-product sink (accumulation of acids) or minimum energy loss via excretion.

Furthermore, Gnaiger (1983c) outlines the selective evolutionary traits in the exploitation of thermodynamically distinctive pathways operating under environmental and physiological anoxia. Maximum rate and maximum efficiency are mutually exclusive. A high rate of ATP demand during exercise is promoted by incurring a necessarily high irreversible energy loss, while an efficient ATP gain as required for anoxic survival prevents the development of a high output power.

9.5 Toxicological and Pharmacological Applications

Heat dissipation has been suggested as a sensitive indicator of the effects of toxicants and pharmaceutical agents on living systems (Gnaiger 1981, 1983b). Antibiotics play an important role in aquaculture of invertebrates and fish, and antibiotics are often used in biochemical, physiological and ecological studies to prevent interference by microbial metabolism, but there is little information of their effect on animal metabolism in spite of their frequent application. Gnaiger (1981) has shown with the aid of calorimetry that animals including aquatic oligochaetes, zooplankton, fish eggs and larvae are influenced to a variable degree by some commonly used bacteriostatic agents, e.g. streptomycin and neomycin sulphate and penicillin G. Streptomycin sulphate (20 mg dm^{-3}) and neomycin sulphate (200 mg dm^{-3}) both induce a lasting increase (>2-fold) in the heat dissipation of the oligochaete, *Lumbriculus variegatus*, in air-saturated water and they exhibit a stimulating effect on anoxic heat dissipation (Gnaiger 1981), although both are known to be inactive against bacteria under anoxia (Dalla Via 1983). In contrast, the combination of streptomycin and penicillin causes a 30% reduction in the aerobic rate of heat dissipation after a transitory increase, but has no effect on the anoxic rate. Furthermore, these antibiotics stimulate the rate and the pattern of fluctuations in the aerobic rate of heat dissipation by zooplankton (*Cyclops abyssorum tatricus*).

The eggs of *Salvelinus alpinus* are not affected by streptomycin and neomycin sulphate, which apparently do not penetrate the egg membrane. After hatching, however, an immediate response to the addition of these antibiotics (200 mg dm^{-3}) is observed as a nearly 3-fold increase in heat dissipation, whereas the oxygen consumption is only doubled (Gnaiger

1983b). This differential response in simultaneous respirometry and calorimetry is due to the activation of anoxic pathways sustaining the high metabolic rate (physiological hypoxia). The anoxic contribution evades detection when only traditional respirometric techniques are applied. Together, these findings suggest that an uncritical application of antibiotics in bioenergetics and physiological studies cannot be recommended (Gnaiger 1981).

Ecotoxicological studies have demonstrated that many environmental contaminants (metals and organics) can have an impact on animal metabolism. However, to date all studies have involved respirometry and both enhanced and depressed respiration rates have been reported, depending on the nature of the toxicant and its concentration. Simultaneous calorimetric and respirometric measurement in an open-flow system now provides the opportunity to gain greater insight into effects of toxicants on the aerobic and anoxic rates of metabolism by aquatic animals.

9.6 Conclusions

Ecological studies involving an energetic approach are able to provide insight into how organisms perform and adapt in response to changes in both natural and anthropogenic environmental factors. Microcalorimetry has played an important role in recent advances in physiological and ecological energetics. The initial application of microcalorimetry was largely of a descriptive and empirical nature, but this has been followed by a more thermodynamic and energetic approach to ecological problems, particularly concerning the metabolic flexibility shown by animals in response to changes in physiological and environmental conditions. To date, a large proportion of these calorimetric studies have involved bivalve molluscs, with particular emphasis on the mussel, *Mytilus edulis*, which serves to illustrate the experimental approach of simultaneous direct and indirect calorimetry and highlight some of the metabolic adaptations of euryoxic animals. The technique promises to make a valuable contribution to the understanding of the complex metabolic responses of a wider range of 'metabolic types', especially when used in combination with biochemical studies, and also in the field of ecotoxicology. Ideally, the metabolic rate should not be measured as an isolated response, but should be related to other principal components of an animal's energy budget. The fundamental processes of energy acquisition and energy expenditure are both responsive to environmental change and ultimately determine the energy available for production. Calorimetry therefore has an essential application in physiological energetics and the construction of energy budgets, which have proved useful measures of an animal's performance and adaptation in response to environmental change (e.g. salinity, Widdows 1985a; air exposure, Widdows and Shick 1985) and the biological impact of pollution (Widdows 1985b).

Acknowledgements

I am grateful to Dr E Gnaiger for useful discussions and for reading the manuscript.

References

Bayne B L and Newell R C (1983) In: Saleuddin A S M, Wilbur K M (eds) *The Mollusca*, Vol 4. New York, Academic Press, pp 407–515

Brett J R and Groves T D D (1979) In: Hoar W S, Randall D J, Brett J R (eds) *Fish Physiology*, Vol. VIII. New York, Academic Press, pp 279–352

Brody S (1945) *Bioenergetics and Growth*. New York, Reinhold, p 1203

Butterworth P E and Barrett J (1985) *Physiol Zool* **58**, 9–17

Crisp D J (1971) In: Holmes N A, McIntyre A D (eds) *Methods for the Study of Marine Benthos*. Oxford, Blackwell, pp 197–323

Curtin N A and Woledge R C (1978) *Physiol Rev* **58**, 690–761

Dalla Via G J (1983) In: Gnaiger E, Forstner H (eds) *Polarographic Oxygen Sensors. Aquatic and Physiological Applications*. Berlin, Springer Verlag, pp 202–18

Ebberink R H M, Zurburg W and Zandee D I (1979) *Mar Biol Lett* **1,** 23–31

Famme P (1981) *Comp Biochem Physiol* **69a**, 243–47

Famme P, Knudsen J and Hansen E S (1981) *Mar Biol Lett* **2**, 345–51

Gäde G (1983) *J Exp Zool* **228**, 415–29

Gnaiger E (1979) *Experientia Suppl* **37**, 155–65

Gnaiger E (1980a) *FEBS Lett* **112**, 239–42

Gnaiger E (1980b) *Thermochim Acta* **40**, 195–223

Gnaiger E (1981) *Thermochim Acta* **49**, 75–85

Gnaiger E (1983a) In: Gnaiger E, Forstner H (eds) *Polarographic Oxygen Sensors. Aquatic and Physiological Applications*. Berlin, Springer Verlag, pp 134–66

Gnaiger E (1983b) *Science Tools* **30**, 21–6

Gnaiger E (1983c) *J Exp Zool* **228**, 471–90

Gnaiger E, Shick J M and Widdows J (1987) (In press)

Hammen C S (1983) *J Exp Zool* **228**, 397–403

Harju-Jeanty P (1982) *Appita* **36,** 26–31

Hemmingsen A M (1960) *Rep Steno Mem Hosp Copenh* **9**, 7–110

Hochachka P W and Somero G N (1984) *Biochemical Adaptation*. New Jersey, Princeton University Press, p 537

Jackson D C (1968) *J Appl Physiol* **24**, 503

Kleiber M (1961) *The Fire of Life*. New York, Wiley, p 454

Knudsen J, Famme P and Hansen E S (1983) *Comp Biochem Physiol* **74A**, 63–6

Koryan S A and Storey K B (1983) *Can J Zool* **61**, 2674–81

Kushmerick M J (1977) *Curr Top Bioenerg* **6**, 1–37

Lasserre P and Tournié T (1984) *J Exp Mar Biol Ecol* **74**, 123–39

Lehninger A L (1965) *Bioenergetics. The Molecular Basis of Biological Energy Transformations*. New York, Benjamin

Livingstone D R (1982) In: Addink A D F, Spronk N (eds) *Exogenous and Endogenous Influences on Metabolic and Neural Control*. Oxford, Pergamon Press, pp 257–74

Ljungholm K, Noren B and Wadsö I (1979) *Oikos* **33**, 15–23

Meyerhof O (1920) *Pflugers Arch* **182**, 284–317

Newell R C (1979) *Biology of Intertidal Animals*. Faversham, Kent, Marine Ecological Surveys Ltd, p 781

Newell R C and Branch G M (1980) *Adv Marine Biol*, vol 17. London, Academic Press, pp 329–96

Pamatmat M M (1977) In: Coull B C (ed) *Ecology of Marine Benthos*. Columbia, University of S. Carolina Press, pp 89–111

Pamatmat M M (1979) *Marine Biol* **53**, 223–9

Pamatmat M M (1980) In: Tenore K R, Coull B C (eds) *Marine Benthic Dynamics*. Columbia, University of S. Carolina Press, pp 69–90

Pamatmat M M (1982) *Science* **215**, 395–7

Pamatmat M M (1983a) In: Gnaiger E, Forstner H (eds) *Polarographic Oxygen Sensors. Aquatic and Physiological Applications*. Berlin, Springer Verlag, pp 167–75

Pamatmat M M (1983b) *J Exp Zool* **228**, 405–13

Pamatmat M M and Findlay S (1983) *Marine Ecol Prog Ser* **11**, 31–8

Pamatmat M M, Graf G, Bengtsson W and Novak C S (1981) *Marine Ecol Prog Ser* **4**, 135–43

Phillipson J (1966) *Ecological Energetics*. London, Arnold, p 59

di Prampero P E (1981) *Rev Physiol Biochem Pharmacol* **89**, 143–222

Prosser C L (1973) *Comparative Animal Physiology*. Philadelphia, Saunders, p 966

Redl B and Tiefenbrunner F (1981) *Water Res* **15**, 87–90

Saz H J (1981) *Ann Rev Physiol* **43**, 323–41

Shick J M (1981) *Marine Biol Lett* **2**, 225–36

Shick J M and Widdows J (1981) *Am Zool* **21**, 983

Shick J M, de Zwaan A and de Bont A M T (1983) *Physiol Zool* **56**, 56–63

Shick J M, Gnaiger E, Widdows J, Bayne B L and de Zwaan (1986). In press

Shoubridge E A and Hochachka P W (1980) *Science* **209**, 308–9

Shumway S E, Scott T M and Shick J M (1983) *J Exp Marine Biol Ecol* **71**, 135–46

Slobodkin L B (1962) *Adv Ecol Res* **1**, 69–101

Sprung M and Widdows J (1986) *Marine Biol* **91**, 41–5

Stebbing A R D (1982) *Sci Tot Environ* **22**, 213–34

Suurkuusk J and Wadsö I (1982) *Chimica Scripta* **20**, 155–163

Taylor A C (1976) *J Marine Biol Assoc UK* **56**, 95–109

Theede H, Ponat A, Hiroki K and Schlieper C (1969) *Marine Biol* **2**, 325–37

Tournié T and Lasserre P (1984) *J Exp Marine Biol Ecol* **74**, 111–21

Wiegert R G (1976) *Ecological Energetics*. Pennsylvania, Dowden, Hutchinson and Ross Inc, p 437

Widdows J (1985a) In: Gray J S, Christiansen M E (eds) *Marine Biology of Polar Regions and Effects of Stress on Marine Organisms*. Chichester, Wiley and Son, pp 555–66

Widdows J (1985b) *Marine Poll Bull* **16**, 129–34

Widdows J and Salkeld P N (1984) *Abstr in 5th Int Symp on Microcalorimetry Applications in Biology*, p. 24

Widdows J and Shick J M (1985) *Marine Biol* **85**, 217–32

Widdows J, Bayne B L, Livingstone D R, Newell R I E and Donkin P (1979) *Comp Biochem Physiol* **62A**, 301–8

Winberg G G (1960) *Transl Ser Fish Res Bd Can* **194**, 1–202

Yetka J E and Wiebe W J (1974) *Appl Microbiol* **28**, 1033–9

Zeuthen E (1947) *C R Lab Carlsb Ser Chim* **26**, 17–161

Zurburg W, de Bont A M T and de Zwaan A (1982) *Mol Physiol* **2**, 135–47
Zurburg W and Ebberink R H M (1981) *Mol Physiol* **1**, 153–64
de Zwaan A (1977) *Oceanogr Marine Biol Ann Rev* **15**, 103–87
de Zwaan A (1983) In: Wilbur K M (ed) *The Mollusca*, Vol 1. New York, Academic Press, pp 137–75
de Zwaan A, de Bont A M T, Zurburg W, Bayne B L and Livingstone D R (1983) *J Comp Physiol* **149**, 557–63
de Zwaan A and Wijsman T C M (1976) *Comp Biochem Physiol* **54B**, 313–24

Index